Science, Life and Christian Belief

A Survey of Contemporary Issues

Malcolm A. Jeeves
and R. J. Berry

A Division of Baker Book House Co.
Grand Rapids, Michigan 49516

41380 466

© 1998 by Malcolm A. Jeeves and R. J. Berry

Published by Baker Books
a division of Baker Book House Company
P.O. Box 6287, Grand Rapids, MI 49516-6287

Printed in the United States of America

First published 1998 by
Inter-Varsity Press
38 De Monfort Street
Leicester LE1 7GP
United Kingdom

Library of Congress Cataloging-in-Publication Data is on file at the Library of Congress, Washington, D.C.

ISBN: 0-8010-2226-6

Chapter five draws substantially upon an unpublished paper, "The God of the physical universe," by David Wilkinson, 1996.

For information about academic books, resources for Christian leaders, and all new releases available from Baker Book House, visit our web site:
http://www.bakerbooks.com

Contents

Preface

A book entitled *The Scientific Enterprise and Christian Faith* was published in 1969. It was the fruit of a small conference in Oxford attended by thirty-six scientists from ten countries, and was written by one of us (Malcolm Jeeves). It was one of the first in-depth publications on the interface of faith and science from a group enthusiastic about the scientific enterprise, comprising Christians who took the Bible seriously. It was widely used by students and ministers and became a course text in a number of universities and colleges.

The first edition has long been out of print and, although there has been a plethora of books on science and faith, it has not been fully replaced by any other work. At a joint meeting of the American Scientific Affiliation and (UK) Christians in Science, a revised version was called for, and the two of us agreed to undertake this. It should be emphasized that we write not as philosophers, historians of science or theologians, but as working scientists.

Some may ask whether we need yet another book on science and religion. We believe there are several reasons. First, most of those currently available deliberately restrict themselves to specific areas of contemporary science. For example, the book by Van Till and his colleagues, *Portraits of Creation*, was deliberately limited. Likewise, the books by John Polkinghorne make little reference to evolution, neuroscience, psychology and issues concerning the environment. This is understandable, since Polkinghorne writes as a distinguished mathematical physicist. Our own books (Malcolm Jeeves, *Human Nature at the Millennium*, and R. J. Berry, *God and Evolution*) are likewise circumscribed. A second reason for yet another book is the rapid increase in the number of college and university courses on science and religion, as well as an increasing number of short courses in theological and Bible colleges. Quite a number of the books currently available are by historians or philosophers of science. They have made important contributions, but it remains the case that there are relatively few written by working research scientists enthusiastic about their science and what it has to offer humankind, and who seek to live as committed Christians. For ourselves, we are determined to avoid the kind of compartmentalization of science and faith into which it is all too easy to slip.

We are, of course, aware of the changes in attitudes towards science and of the interpretations of Scripture evident over recent decades. Initially we naïvely expected that the earlier book would require little amendment, and that our task would be to revise the later chapters in the light of advances in genetics and neurology, of our understanding of behaviour, and of the escalating results of irresponsibility towards the environment. We soon found, however, that the whole book needed revising. Part of the reason for the rewrite was that the first edition betrayed signs of its conference roots. We have sought to remove these. Much more important have been advances in theological interpretation and the public understanding of science. As science has expanded, older attempts to 'find' God in his works have been radically affected, validating ever more strongly the complementary model of divine action expounded in the first edition. In addition, concern has shifted from academic problems over physical determinism to the practical ethics of psychological and genetic determinism. We have tried to include these recent changes within a coherent picture of a Universe created and upheld by a God who seeks to make himself known. We fully accept the revelation of God in his written and living Word; the problem in every generation is to interpret this revelation in a consistent way. This has sent us back repeatedly to check to the best of our understanding what the Bible *actually* says, as distinct from how it has been *conventionally interpreted*. We have been repeatedly reminded of Galileo's response to Pope Urban VIII, who wanted Galileo to agree that Scripture taught a fixed Earth, with planets and Sun circling round it. Galileo made the reasonable and vivid point that the Bible 'teaches us how to go to heaven, not how the heavens go'. Our approach is the same.

Two of the most seminal contributors to the science–faith debate in recent years have been Reijer Hooykaas and Donald MacKay. Both were at the Oxford meeting which inspired this book and at the ASA-CiS meeting twenty years later; sadly, both have since died. As individuals and as authors we are indebted to them. We are also grateful to many others who have read and commented on our manuscript. Our thanks are especially due to John Hedley Brooke, Paul Helm, Colin Russell and Harry Hine (chapters 1 and 2); Howard Van Till and Roger Trigg (chapters 3 and 4); David Wilkinson and Arnold Wolfendale (chapter 5); Ernest Lucas (chapters 6 and 7); Richard Bauckham (chapter 8); Duncan Vere (chapter 9); Anthony Thiselton (chapters 8 and 9); Warren Brown and David Myers (chapters 10 and 11); Ghillean Prance (chapter 12); John Houghton (chapters 5 and 13); and particularly to Oliver Barclay, Richard Bube and Allan Day for reading the whole manuscript. We have tried to take into account all their corrections and suggestions, but the responsibility for the overall interpretations must remain with us.

One of the major challenges in writing a book such as this is how to communicate effectively with both the non-scientist and the scientist. The attempt to do this has to be made if one of the principal aims of the book is to be achieved, namely, to set down with supporting arguments why we believe that science is a true friend of biblical faith and not, as is often asserted, in conflict with it. While all Christians must be conscious of the impact of science on their daily lives and faith, there are two particular groups who are probably more consistently aware of this impact than others: first, thoughtful students reading for a first or postgraduate degree in science, and second, leaders of Christian groups, themselves often non-scientists. Ministers in local churches, for example, are increasingly, and at times embarrassingly, conscious of their lack of familiarity with basic scientific ideas. Young people in their congregations often pose problems which they cannot understand, let alone begin to answer. Such leaders may then struggle with specific issues arising at contact points between particular religious beliefs and specific scientific discoveries, dealing with them either on a purely *ad hoc* basis, or within an inadequate framework of the biblical view of nature or of the scientific enterprise, or both.

Since the earlier book, *The Scientific Enterprise and Christian Faith*, was both out of date and out of print, another reason to attempt a revised and expanded update was the need to address *contemporary* issues at the science and faith interface, concentrating on natural science, apart from a brief treatment of social psychology. Of today's problems, while some are new, many are not. Probably most are old ones in new dress and accordingly are best understood, we believe, when seen in their historical context. Thus, chapter 1 offers a brief historical sketch with references for further reading, for those interested in following the matter in more detail. We then move on to three key sets of concepts, the understanding of which sets the framework for considering specific issues. First, for those, like us, who take the biblical teaching seriously, we need to ask what guidance we find therein concerning the relation of God to his creation. Next, we need to ask what is meant by *the* scientific enterprise and *the* scientific method. What is distinctive about it, as a route to increasing our knowledge of ourselves and the world we live in? That in turn requires us to look more closely at what we mean when we say we can explain some aspect of the created order in scientific terms. What is the nature of the models we construct and find useful in science? We need to ask this because there are other ways of giving an account of, for example, a sunrise or a sunset than those given by scientists. Are they to be seen as competitors with the scientific accounts? If the answer is no, then how should we think about them in relation to the scientific accounts? From there onwards, we move systematically through a discussion of issues that arise in the physical sciences (chapter 5), biological sciences (chapters 6 and 9) and the psychological and neurosciences (chapters 10 and 11), pausing from time to time to enquire what relevant biblical teaching there may be on some of the issues discussed (chapters 7 and 8). Finally, we look ahead in two senses. First, we ask what we are doing and what we should be doing about our stewardship of creation (chapter 12), and secondly, we consider what are the principles that should guide our practice as Christian scientists (chapter 13).

Since, as we have indicated, we wish this book to be helpful to those who organize and contribute to courses in the general area of science and religion, we have included numerous quotations from and references to source materials. In this respect the book departs from a typical book for the general reader, though we hope they will not be put off by the additional documentation, most of which we have put in the endnotes. To assist the reader further, we give a brief summary, at the start of each chapter, indicating the questions addressed and the arguments offered. We hope this will help to explain why the issues discussed have been selected.

Malcolm Jeeves
R. J. Berry

Chapter 1

Hebrew–Christian and Greek influences on the rise of modern science

For more than two millennia, wise thinkers have sought to understand themselves and the world they live in. The Greeks, along with other ancient civilizations, saw the world as controlled by indwelling, divine forms. For the ancient Hebrews, the world was non-eternal, created and upheld by God. For them, God and nature must be sharply separated. God is eternal and changeless; creation will one day pass away. Over the centuries, the interactions between Greek and Hebrew-Christian views of creation make up a complex story. Awareness of it helps to put current issues at the interface of science and religion into a proper perspective. It also helps to avoid repeating past errors. One such issue given much publicity today is the emergence, decline and re-emergence of creationism. It benefits from being put into a proper historical perspective.

The Greeks, along with all their magnificent intellectual and cultural achievements, 'possessed an elaborate scientific world picture and laid the foundations of some scientific disciplines, such as astronomy and optics' (Hooykaas 1971: xii). Why did they fail to sustain the rise of science two thousand years ago? Why did the scientific revolution not begin in earnest until the sixteenth century, and then flourish? Can its rise be related to that other re-awakening which began in the sixteenth century, namely the religious Reformation? Answers to these questions underpin many of the current issues at the science–faith interface today.[1]

Attitudes towards the natural world began to develop during late medieval times and set the stage for the emergence of modern science. The two major influences were Greek and Hebrew-Christian. These two streams have flowed together for many centuries, interacted in numerous ways, and produced reciprocal alterations in each. In these two streams were the salient ideas which were to be essential to the rise and development of modern science, as well as other ideas which delayed the beginning of modern science as well as threatening its continuation and development.

Greek influences

It is misleading to speak of *a* Greek view of nature, in that it telescopes a wide spectrum of views which we now know were held by Greek thinkers in the six centuries up to the third century BC, when the centre of intellectual enquiry moved from Greece to Alexandria (see below, p. 19). Notwithstanding, there are certain themes which recur in earlier times which can properly be described as Greek views. These have had an enduring influence on all later western thought.

Greek thinkers firmly held that the Universe should be regarded as existing from eternity to eternity and therefore as non-created. Guthrie puts it thus: 'The most noteworthy believer in the eternity of the universe was, of course, Aristotle, and he makes frequent mention of the cyclic theory of human affairs . . . But it occurs in his master Plato, and also in later writers like Polybius and Lucretius' (Guthrie 1957: 65). As well as being eternal, the Universe was often regarded as divine or semi-divine. Behind it was 'the same mysterious *x* which is what we might call the material sub-stratum of the world, [and] is also the force which guides or directs it. It is not only everlasting, but everlastingly alive, immortal and divine' (1957: 48). This was broadly true of most of the ancient world, including Assyria and Egypt. The Greeks considered that since the Universe was regarded as divine, the world was being moved in a purposeful way by divine forces. Guthrie, while accepting that our knowledge of the early Greek philosophers is limited by the loss of their writings, nevertheless insists that we can discern the main elements of their thinking, and 'there is no need for scepticism . . . when we read that Anaximenes thought of the air as God, and also drew an analogy between the air which sustains the universe and the human soul. The idea that the whole world is a living and breathing creature was firmly upheld by the Pythagoreans and finds its most striking expression in Timaeus . . . Thus everything is made of one substance, and that substance, at least in its most properly balanced, invisible form, is the substance of life. Since it is everlastingly alive, it is divine, for immortality and divinity were two inseparable concepts for a Greek. The life principle in finite creatures is the same. Perhaps that is what Thales meant in one of the few sayings which can be plausibly assigned to him: "everything is full of Gods"' (1957: 49). More recent scholars, for example Barnes (1979: 1:94–99; 2:279–280), tend to be more cautious about attributing pantheistic views to pre-Socratic philosophers. On the other hand, from the third century BC onwards, the Stoics maintained that the Universe is permeated and controlled by a rational, divine force (Sandbach 1975: 69–82).

This indwelling of nature by divine forces provided a direct link between humankind and nature, since the human mind was also believed to possess the same divine character attributed to the forms of nature. Such a belief did not go unchallenged, since the doctrine of Diogenes of Apollonia that 'the air within us is a small portion of the God' was parodied by Aristophanes in one of his comedies, when he brings Socrates on to the stage suspended in the air in a basket. Asked the reason for the strange proceeding, Socrates replied that 'to discover the truth about celestial matters, he must allow his mind to mingle with the kindred air' (Guthrie 1957: 50).

Since both humankind and nature were alike divine, it was assumed that the one could read off or intuit the other. And because, for the Greeks, reason was the principal tool of the human mind, reason must be the key to the mysteries of nature. By 'reason' they had in mind not primarily rational thought but *a priori* deduction from certain axioms. It was a firm belief in reason that was the cornerstone of Greek speculation about the origin and nature of the Universe. Kitto reminds us: 'Here we meet a permanent feature of Greek thought: both the physical and the moral universe must be not only rational, and therefore knowable, but also simple' (Kitto 1951: 179). So great was this belief in reason that the Greek 'tended to impose pattern where it was in fact not to be found, just as he relied on reason where he would have been better advised to use observation and deduction' (1951: 187).

It is here that we can glimpse ways in which the tremendous achievements of the Greeks in the development of logical reasoning became the very factors which placed limits on the development of empirical science: 'The ancient Greek believed fundamentally that the world should be *understood*, but that there was no need to *change* it. This remained the belief of subsequent generations up to the Renaissance. This passive attitude to the practical use of the forces of nature was reinforced by the complete ossification of the natural sciences in the Middle Ages in the condition to which Aristotle had brought them' (Sambursky 1963: 230). The difference nowadays is that 'In modern science there is a balanced use of induction and deduction; theory and practical application help one another. The Greeks, on the other hand . . . formed *a priori* hypotheses too readily, and neglected close and systematic observation and experiment – largely because their main motive, as Aristotle approvingly said, was curiosity, rather than the attempt to dominate nature' (Kirk 1958: 111). Today, these views are more nuanced to take account, for example, of the recognition that Greek medicine *was* trying to *change* people's state of health. Moreover, while there was a neglect of experiment, there was not a total absence. Furthermore, areas of science of great interest to the Greeks, such as astronomy, were not at that time readily susceptible to experiment. There was certainly plenty of *observation*, notably in astronomy and medicine.[2]

We thus see that the magnificent intellectual achievements of the Greeks, which guided western thought for two millennia, had inherent limitations which prevented them from bringing forth modern science. We can identify three features of Greek thought which most consistently inhibited the development of a modern scientific approach. In the first place there was no necessity for empirical testing, since, as we have seen, 'the ancient Greek believed fundamentally that the world should be *understood*, but that there was no need to *change* it'. It was unnecessary to stoop to handle the world in order to understand it; contemplation allied to reasoning was sufficient. Secondly, it was perfectly natural to look for teleological explanations to the exclusion of other types. And thirdly, because the human mind was rational, the Greek elevated intuition and the use of reason above careful observation: 'the philosophers tried to explain nature while shutting their eyes' (Guthrie 1953: 190).

Despite these shortcomings, there were features of Greek thinking which were germs for the subsequent development of modern science. In the first place, the world *did* have order in spite of its apparent chaos: 'The Greek never doubted for a moment that the universe is not capricious: it obeys Law and is therefore capable of explanation' (Kitto 1951: 176). Secondly, there was clearly value in a body of knowledge about nature. Thirdly, the Greeks discovered and perfected a method of deductive reasoning. Fourthly, they produced some remarkable developments in philosophy, logic and mathematics, all of which were to be basic and essential tools for the eventual rapid development of science. In the same way that 'their eyeshutting retarded the growth of science', so 'their opening led to things perhaps equally important, metaphysics and mathematics'. And finally, they made significant advances in astronomy, physics (mechanics) and biology (medicine).

In short, the main legacy of the Greeks in the development of science is to be found, not in particular theories, but in their general attitude of rational investigation of nature by means of logic, mathematics and observation, a task already well in hand by the Egyptians some time before. Although they may not have felt any necessity for empirical testing which involved *changing* nature, some of them emphasized that aspect of the empirical method in which *observation* plays a dominant role 'so that even though Greek influence on science may not seem as monolithic as is often imagined, the very diversity of its effects testifies to its immense significance for the emergence of science' (C. A. Russell 1985: 29). Such a view is reinforced by noting how some of the great heroes of the scientific Renaissance expressed a profound debt to their Greek precursors, as Galileo did to Archimedes, Copernicus to Pythagoras, and Harvey to the Aristotelian philosophy he encountered at Padua.

Reciprocal influences on Greek science

Greek science flourished for more than a thousand years. While all attempts at classification are somewhat artificial, we may note four main periods. (1) From 600 to 420 BC was the age of the pre-Socratic philosophers. (2) From 420 to 220 BC the key figures were Hippocrates, Plato, Aristotle and Archimedes. From 200 BC to AD 50 there was a gap in progress (apart from Hipparchus, a major figure in astronomy), but then (3) in AD 50 to 200 another major flow of writings on science took place. (4) The final period extends from AD 370 to the Arab Conquest. It was the scientists of this late phase of late antiquity who were the conduit for the Arabs and then the Middle Ages. 'Science in the Greek world was not a unity but a plurality. It was not a single way of viewing the world, but one that was complex and at times self-contradictory' (Nutton 1993). Perhaps the most important facet of Greek science was that it was open, visible and accessible to all, not a secret gnostic cult. In the modern sense, 'Ancient science was public science, for both good and ill' (Nutton 1993).

This is obviously a generalization; some forms of Greek science were open, but others were not. For example, the Pythagorean brotherhood was rather closed and élitist, cultivating a sense of the esoteric. Indeed, centuries later Isaac Newton believed that some of their precious 'secrets' had been lost.

Between 320 and 220 BC, Alexandria became the intellectual capital of the Hellenistic world (Lloyd 1973; 1991; Von Staden 1989). It was there around 280 BC that the human body was dissected for the first time. On Alexander's death, Egypt was seized by Ptolemy, who established a dynasty that lasted until the Roman conquest of Egypt and the death of Cleopatra and her son in 30 BC.

Much Alexandrian science survived the upheavals of the period of the later Ptolemies, notably the actions of Ptolemy Euergetes III, who was said to have filled the towns and islands of Greece with grammarians, philosophers, mathematicians, musicians, painters, teachers, doctors and other craftsmen by expelling them from Alexandria because they had supported his brother against him in a civil war.

Colin Russell, typical of his fellow historians of science, has underlined the importance of the temple of the Muse (or Museum) established by the first Ptolemy as a centre of both literary and scientific research, but in AD 390 the Alexandrian museum was partially destroyed by Bishop Theophilus.

Nevertheless, Alexandrian science survived and was still being taught the following century in Syria. It was, however, the rise of Islam in the seventh century AD that introduced a totally new factor. The influence of the prophet Muhammad, born AD 570, spread rapidly through the Near East. With it came a new Arab unity, and the centres of Greek culture were overwhelmed. The Arabs took Jerusalem and Alexandria, and science became the property of the Muslims. The result was that much of Greek culture was available in the West only through devious contacts via Spain and Sicily. Greek texts available in medieval Christendom were better described as 'perversions' than 'versions', having been translated from Greek to Syriac to Arabic to Spanish and then to Latin (C. A. Russell 1985: 26).

Views about the contribution of Greek thought to the intellectual and cultural explosion of the Renaissance have been repeatedly revised. A hundred years ago it was generally believed that Renaissance science was wholly dependent upon its Greek origins. On this interpretation the scientific revolution was understood as a product of Greek thought banishing the mists of medieval theology: rationality overcame superstition; science had won round one of its fight with religion. Few historians now accept this conclusion. The modern view incorporates the importance of Hebrew-Christian roots as an integral part of science's heritage: 'Science arose in the West, not when Christian theology was submerged by Greek rationalism, but rather when Greek and other "pagan" ideas of nature were shown to be inadequate in the new climate of biblical awareness brought about by the Reformation' (C. A. Russell 1985: 55).

Chronologically, of course, the Old Testament pre-dates the intellectual flowering of both the ancient Orient and Greece. But it was only at the Reformation, with the new availability of the Bible to the masses and its re-emergence as a major cultural force, that it added impetus to the development of science. This new energy, flowing together with all that was best in Greek thinking and eliminating some of its worst features, detonated a chain reaction leading to the exponential growth of science from the sixteenth century to the present day.

Just as scientific progress was given its initial impetus by the Greeks, so it was eventually inhibited in Greek culture by the same *a priori* attitude which began it. There is an enormous irony in this: a highly developed system of deductive reasoning stifling new discovery. We must now turn to see how it was changed by the penetration of ideas coming from other traditions, notably the Hebrew-Christian tradition.

The Hebrew-Christian view of nature

What are the salient features of the Hebrew-Christian view of nature? The most outstanding one is that of a world totally dependent upon God.

Consider some implications of this. By contrast with the Greek view, we find that the world is non-eternal, and that it is created by God (Gn. 1) and dependent upon him for its continuing moment-by-moment existence (Heb. 1:3). God and nature are not to be identified with each other; they are rather to be sharply separated. God is eternal, and nature is created and will one day pass away. As the psalmist put it,

> In the beginning you laid the foundations of the earth,
>> and the heavens are the work of your hands.
> They will perish, but you remain;
>> they will all wear out like a garment.
> Like clothing you will change them
>> and they will be discarded.
> But you remain the same,
>> and your years will never end (Ps. 102:25–27).

Secondly, the natural order is not divine; it is created. *God* made the firmament (Gn. 1:7). He ordered it; it is not autonomous. 'The moon marks off the seasons; and the sun knows when to go down. You bring darkness, it becomes night' (Ps. 104:19–20). The createdness of the world is emphasized at the opening of John's gospel, for 'Through him all things were *made*; without him nothing was made that has been made' (Jn. 1:2). Not only did God take the initiative in creating the world, but the world continues to exist only by reason of his activity in 'sustaining all things by his powerful word', as the author of the letter to the Hebrews puts it, writing of the work of Christ (Heb. 1:3). For the Greeks, the workings of nature were detectable by reason alone, with the purposefulness embedded within nature itself; in the Hebrew-Christian tradition, purpose resides in God and not in nature.

For the Hebrews, God alone is to be worshipped; nature is his creation and to worship it is idolatry – and the Lord God tolerates no idolatry. As Deuteronomy 6:4 teaches, 'The LORD our God, the LORD is one', and in Genesis 1:28, we are to 'fill the earth and subdue it', we are to 'rule [or have dominion] over' it. Creation is a gift and a trust from God; it is not God himself. Since our mind is part the created order, it is non-divine and subject to error; it cannot infallibly read off from nature the inherent qualities of nature. In any case, there are firm limits to our intellectual understanding, since God's thoughts are higher than our thoughts (*cf.* Is. 55:8). If we wish to discover the patterns of order in nature we must have recourse to experiment and experience; intuition or reason alone is insufficient.

How do these features of the Greek and Hebrew-Christian attitudes towards nature affect the rise of science?

Interacting streams and the rise of science

Before the seventeenth century, there were brief occasions when the Greek and Hebrew-Christian streams interacted, notably from the eleventh century onwards, when Aristotelian natural philosophy was introduced into the West, and Thomas Aquinas brilliantly synthesized it with Christian thought. We thus find scholastic thought, based upon reason and revelation, linking elements of Greek philosophy and biblical faith. For Aquinas, the crucial feature of all events in the world was their contingency; they might have not occurred, but in fact they did. Aquinas portrayed God not just as the original creator, but as the one who continues to rule over nature and to sustain it from moment to moment. At the same time he accepted many of the deductions from Aristotelian philosophy, such as that it was *a priori* impossible that natural compounds could be made by human beings, and that it was impossible that heavenly bodies should have any motion except perfectly circular ones. He vigorously defended the miraculous, and in fact encouraged the study of nature (in Aristotelian terms), so that miracles might not be ascribed to other causes. According to Aquinas, knowledge of natural things enables us better to recognize supernatural events. This sharp distinction between natural and supernatural events looms large in much subsequent Christian thought.

But it did not go unchallenged. For example, in 1277 the Bishop of Paris (Etienne Tempier) condemned anything that submitted God's incomprehensible will to human reason, and hence he allowed possibilities excluded by Aristotelian philosophy, since he believed that human beings cannot decide beforehand what it pleases God to perform in nature (Hooykaas 1957: 16).

Similar protests against Greek and scholastic rationalism recurred throughout the next century, in particular when the Nominalist philosophers asserted that nature should be accepted as readily when it transcends, or seems to contradict, human reason as when it does not. 'Precisely because they discarded rationalistic pretensions, the Nominalists could tackle scientific problems in a rational way; their reason was not their god; it was used to criticize rather than to erect deductive systems. They realized that science does not result in absolute truth, but is a human methodical approach to divine "revelation" in nature' (Hooykaas 1957: 17).

It was Galileo who showed that a rational approach, properly used in conjunction with experimentation, could lead to significant scientific advances. The way in which theory combined with experiment yields essential ingredients for the development of science was worked out in Galileo's careful investigations. Other features of Galileo's thought were not so commendable, for with him began the development of a view whereby

God became little more than the original creator of the interacting elements. Nature, once created, was considered to be independent and self-sufficient. This trend became dominant in the seventeenth century as the medieval and Reformation emphasis on God's direct and active relationship with his world declined. It is not surprising that there was fertile ground for Cartesian dualism, emphasizing that the sequence of events in the world is determined by mechanical law and not divine action. But even that interpretation needs qualification. Descartes actively affirmed God's continual role in sustaining and conserving motion in the world. He even spoke of God's action at each moment, as if he were constantly recreating the world. On the other hand, Descartes's aim to explain miracles in naturalistic terms and his account of the development of our solar system understandably made no reference to divine intervention or control. This meant that he was assumed to be implying deism (that is, God is effectively an absentee landlord), even if that was not strictly his position. There was clearly a wide range of metaphysical beliefs among Catholic scientists, with Galileo, Descartes, Gassendi and Mersenne among the early mechanists.

If the mathematical approach to nature is displayed by Galileo and Kepler, the experimental tradition finds its champions in people like Francis Bacon. To Baconians, nature was not 'logical' and 'rational', but 'given'. According to Bacon, we forfeit our dominion over nature by wanting to make her conform to our rationalistic prejudices, instead of adapting our conceptions to the data of observation and experiment. Hooykaas has contrasted this aspect of Bacon's thought with that of Simon Grynaeus of Basel (1550), a friend of Calvin:

> Grynaeus, as well as Bacon, regarded the discovery of new parts of the earth as evidence of the restoration of our dominion over nature, but Bacon stresses the submission to facts, however *unexpected* they may be, whereas Grynaeus passes by this humiliating situation and glories in the fact that the human mind, by means of mathematics and astronomy, had in a certain respect anticipated these discoveries and travelled through the universe without needing immediate observation. He merely exults in the strength of human reason and laments, not that we overestimated his faculty, but that we left it unused.
>
> Bacon learned the lesson that we should 'seek for the sciences not arrogantly in the little cells of human wit, but with reverence in the greater world'. He expects the restoration of science to come by the liberation of the mind 'from the serpent's venom that made it swell', and by 'true humiliation of the spirit'. Grynaeus, on the contrary, only finds opportunity to boast of the acuteness of human wit 'which transcends the forces of nature' (Hooykaas 1957: 19).[3]

Enlightenment and later

In England, the seventeenth century was a time when many Puritan scientists welcomed science as an ally of true religion, and were protagonists for a free science, unconstrained by the superstitious cult of Aristotle. The interesting point of this assertion is the danger they perceived in appealing to any authority other than that of revelation, so their free science was 'not adorned by great names, but naked and simple'.[4]

With the Enlightenment we meet its great prophet, Sir Isaac Newton, adhering to Baconian empiricist ideals in the face of rationalism and deism. Thus in 1713 he argues against Cartesian necessitarianism on the same grounds on which Tempier in 1277 had condemned that of the Greeks and even of Aquinas. The same was true of the majority of eighteenth-century philosophers and scientists, accepting as they did the necessity for empiricism, yet at the same time the danger of submitting its fruits to the rationalistic spirit of the age and thus forfeiting something of the true freedom of science. Perhaps the best way of regarding the Enlightenment is as a secularized puritanism; the essential difference between the eighteenth-century and the Puritan Enlightenments is that to the Puritans it was not freedom which led to truth, but truth which led to freedom. This is a crucial distinction. For those for whom the eighteenth century was an age of optimism, it was natural to develop an excessive faith in reason, qualified for some with the reservation that one's views on larger issues had to be constantly tested at the bar of revelation. The Puritan view of fallen humanity constrained them to develop a realistic assessment of reason, which required aid either by revelation in matters of faith or by observation and experiment in matters of science.

Later in the eighteenth century and early in the nineteenth, the Romantic poets reasserted God's immanence in his world. For them, nature was not simply the raw material for scientific analysis; it contained another aspect, beauty, which spoke of a deeper spiritual reality. In the face of Laplace's perhaps exaggerated emphasis on the over-against-Godness of nature, leading to a deterministic and reductionist view of nature as a self-sufficient machine, it is not surprising to see the Romantics reacting with a view verging on pantheism.

This spirit of optimism showed itself also in a desire to extend the methods of science to all problems and all realms of knowledge. Typical of this is Hume's view that an idea which cannot be traced to specific sense data is without significance – a true forerunner of the positivism of the 1920s (see p. 52). He was, of course, challenged, for we find Kant strongly asserting the crucial part played by the human mind in *supplying* the categories for grouping and interpreting the empirical data of science, rather than waiting for them to arise in some unspecified manner from the data themselves.

Moving into the later nineteenth century, yet another fairly widely accepted product of Aristotelian doctrine came under scrutiny. It had seemed natural, *reasonable*, to conclude from Aristotle's teaching that, since all living things are in some sense embodiments of eternal forms or unchanging essences, species are therefore also fixed and unchanging. With Darwin, Aristotelian biology was shaken at its very foundations. What if species are not fixed? What if there is a measure of change from generation to generation? What Galileo had done to Aristotelian physics, Darwin did to Aristotelian biology; Darwin, like Newton, was the point of departure for a new worldview. Newton's intelligently designed machine under Darwin's influence acquired the properties of a dynamic and progressive process.

At the same time, and perhaps strangely, we find aspects of Darwin's views recapturing a Hebrew-Christian emphasis on human nature. Nature, said Darwin, *includes* human beings and culture together. By contrast, the Greeks separated us from the rest of creation and so gave us and our mind an arrogant, aristocratic place against nature. Darwinian views also challenged again any simple analogy of God as the 'Maker' of the universe, that is as an absentee landlord who had made the world and then left it to run autonomously.

We return to this point in chapters 6 and 12. For our present purposes, however, we note particularly how Darwin's views raised in an acute form the issue of what sort of conclusion can properly be drawn from the methods of science. Can the idea of progress, for example, as championed by so many in the nineteenth century, be derived from science? Can ethical norms be derived from a scientific theory? T. H. Huxley, for example, answered, 'No'; Herbert Spencer thought 'Yes'. Sides were taken on this latter issue, both then and since. Another issue, which we shall take up in detail later when we discuss the 'God of the gaps' (chapter 4), was sharpened up at this time. For as Darwin and those who followed him provided possible scientific explanations for features of the natural world hitherto not understood scientifically, so the places where God was still considered to act were slowly eliminated. And all this had the wholesome effect of forcing a new look at the relation of God to his creation. Furthermore, since the scientific evidence for evolution was claimed variously to support naturalism, evolutionism or theism, questions concerning the nature of the scientific enterprise were raised with a fresh urgency. Could one, for example, legitimately extrapolate scientific theory into a metaphysic, or not? Clearly, some thought 'Yes', others thought 'No'. It is to these two recurring questions, of how we should understand the relationship of God to his creation, and how we should understand the scientific enterprise and the knowledge to which it gives rise, that we shall turn in the next two chapters.

The preceding all-too-brief overview would need qualification in a more extended treatment. Specifically, for example, Laplace's alleged reductionism was almost certainly unknown to Blake, Wordsworth, Davy and Coleridge. It was not solely the realm of ideas that was the formative influence at this time. The profoundest social changes were occurring: war, industrialization, political movements in which natural theology was enrolled to aid in social control. All fed into the overall changes which were occurring in ways of thinking about life in general and the way the world is in particular.

Science and faith: friends or foes?

Since Christians were so influential in the development of science, and since Hebrew-Christian thinking about the natural world positively facilitated the rise of science, it is natural to ask how the popular view of the relation between science and religion as conflict, rather than as mutual support and alliance in a common cause, came about. The conflict metaphor is certainly a widely held myth, and needs demythologizing as urgently as other myths so beloved by some theologians.

There is a widespread assumption that there has been a deep and enduring hostility between science and religion. It is repeatedly popularized in the media. Colin Russell cites the influential *The Ascent of Man* by Jacob Bronowski (1974), but it is just as likely to surface in letters to the broadsheet press over issues such as the virgin birth, resurrection, and miracles. When the first edition of this book was written, Malcolm Dixon and his student John Habgood, later Archbishop of York, could both refer to the 'uneasy truce' between science and faith (C. A. Russell 1989: 5). Russell believes that to portray Christian belief and scientific knowledge as persistently in conflict is 'not only historically inaccurate but actually a caricature so grotesque that what needs to be explained is how it could possibly have achieved a degree of respectability' (C. A. Russell 1989: 5–6).[5]

Evolution and Christianity

A prime example of such a conflict is the continuing debate among some Christians about evolution. It was natural for pre-nineteenth century thinkers who today would be called scientists to believe in special creation. Both the Bible and current philosophical presuppositions pointed to it. But as knowledge of fossils, geological processes, extinctions, and the distribution of animals and plants increased, so the traditional interpretation became more strained, and additional explanations had to be added.

Darwin's theory and marshalling of evidence in many ways released the tensions that had been building up over the understanding of the natural world, and his ideas were fairly rapidly accepted by the majority of both the scientific and Christian worlds. There were, of course, exceptions. One of the most important Christian opponents was Charles Hodge of Princeton, who coined the aphorism: 'What is Darwinism? It is Atheism.' Hodge was not against evolution or natural selection as such, however, but against their effect on divinely controlled purposes:

> It is neither evolution nor natural selection which gives Darwinism its peculiar character and importance. It is that Darwin rejects all teleology, or the doctrine of final causes. He denies design in any of the organisms in the vegetable or animal world . . . and it is this feature of his system which brings it into conflict not only with Christianity, but with the fundamental principles of nature religion, it should be clearly established (Hodge 1874: 48, 52).

There is no doubt that Hodge's views should be accepted as fuelling mainstream 'creationism', but we need to be careful about too comprehensively claiming him as the patron saint of anti-evolutionism. In 1868 a Scottish Presbyterian minister, James McCosh, was appointed President of Princeton University (then known as the College of New Jersey). McCosh's reputation was based on a book, *The Method of Divine Government* (1850), which taught that God governs the world both by law and through a 'complication' of laws which produce 'fortuities', an interpretation not too far removed from what has come to be called 'complementarity' (see chapter 2). In his first week at Princeton he told the senior students that he was fully in favour of evolution provided that it was 'properly limited and explained'. Later, he reminisced about his time at Princeton: 'I have been defending Evolution, but, in doing so, have given the proper account of it as the method of God's procedure and find that when so understood it is in no way inconsistent with Scripture . . . We give to science the things that belong to science, and to God the things that are God's. When a scientific theory is brought before us, our first enquiry is not whether it is consistent with religion, but whether it is true' (McCosh 1896: 184, 234).

The point of introducing McCosh here is not to indicate that Christians expounded evolutionary ideas in the early years after the publication of the *Origin of Species*, but to record that Charles Hodge warmly welcomed McCosh to Princeton, and during the inaugural ceremony declared that never in the history of the college had an academic appointment received such universal approbation.

'Creationism' and fundamentalism

Nowadays, 'creationism' is taken as almost synonymous with 'fundamental-
ism', but again we must beware about tracing modern-day 'creationism' to
the original fundamentalists. The first fundamentalists were named on the
basis of 'five fundamentals' drawn up by the 1910 General Assembly of the
American Presbyterian Church, which were intended to represent
the fundamental beliefs of Protestant Christianity: the miracles of Christ;
his virgin birth; his sacrifice on the cross constituting atonement for
humankind's sin; his bodily resurrection; and the Bible as the directly inspired
Word of God. These 'fundamentals' were expanded in twelve booklets
published between 1910 and 1915, which included an essay by James Orr,
Professor of Systematic Theology in the Glasgow college of the United Free
Church of Scotland. He argued that the Bible is not a textbook of science;
that its intent is not to disclose scientific truth but to reveal the will and
purpose of God; that the world is 'immensely older than 6,000 years'; that
the first chapter of Genesis is a 'sublime poem' which science 'does nothing
to subvert'; and although evolution is not yet *proved*, there seems to be a
growing appreciation of the strength of evidence for some form of
evolutionary origin of species (Orr 1910: 103). Another contributor, George
Wright, a distinguished geologist, declared forcefully that 'if it should be
proved that species have developed from others of a lower order as varieties
are supposed to have done, it would strengthen rather than weaken the
standard argument from design' (G. F. Wright 1910). Yet another contributor
was the Princeton theologian B. B. Warfield, well known as an authoritative
defender of the authority and inerrancy of the Bible. Warfield believed that
evolution could supply a tenable 'theory of the method of divine providence'
in the creation of humankind. He took pleasure in showing that Calvin's
doctrine of the creation, 'including the origination of all forms of life,
vegetable and animal alike, including doubtless the bodily form of man', was
a 'very pure evolutionary scheme' (Warfield 1915: 196).

Clearly, the fundamentalists *sensu stricto* cannot be claimed as the
progenitors of 'creationism'. Whence, therefore, comes the movement? One
of the clues must be the so-called 'Tennessee Monkey Trial' of 1925, when
John Scopes, a local schoolteacher in Dayton, was convicted and fined $100
for violating a law forbidding the teaching of evolution. (The sentence was
later overturned on a technicality by the state appeal court because the judge
had set the fine, rather than the jury, as the law required.)[6]

The issues were sociological rather than theological. In the 1920s, America
was changing from a nation of farmers into a nation of city-dwellers and
wage-earners. The watershed of the First World War had reduced the old
empires, opening up an avenue of money and power into which much of

America stepped enthusiastically. As the only industrial nation not devastated by the war, America was suddenly *the* world power. As a consequence, rural America became out-voted, out-shouted and out-financed; the farmers saw themselves relegated to pawns in battles between railroad barons, bankers and industrialists. Scopes's prosecutor, William Jennings Bryan, was their hero (p. 256). Rural populism developed, preaching a return to the simple days of yeoman farmer and the craftsman. The populists were thus a revitalization movement harking back to a golden past, 'anti-intellectual' in effect, because they opposed 'progress' that was rending apart their previously comfortable (or at least familiar) social and economic fabric. Industrial technology was the demon; educated people invented and ran it, and evolutionists, the populists thought, defended it. 'Too much' education was seen as wasteful and morally questionable, except in practical fields like medicine, agriculture or law. By the time of the Scopes trial, populism was largely restricted to religion where God gave comfort as the political and economic system had not. From political revival with religious overtones sprang religious revival with political overtones, its secular power dormant.

'Bible-science'

In 1941 the American Scientific Affiliation was formed to explore the relationship between science and the Christian faith. Initially it supported a literal interpretation of the Genesis creation accounts, but it gradually moved to a less literal (albeit no less avowedly Christian) stance. Then in 1961 a major event in 'creationist' history took place with the publication of *The Genesis Flood*,[7] written by a theologian, John Whitcomb, and a hydraulic engineer, Henry Morris. It was explicitly apologetic. The authors wrote: 'We believe that most of the difficulties associated with the Biblical record of the Flood are basically religious, rather than scientific. The concept of such a universal judgment on man's sin and rebellion, warning as it does of another greater judgment yet to come, is profoundly offensive to the intellectual and moral pride of modern man and so he would circumvent it if at all possible' (Whitcomb & Morris 1961).

'Creationism' in Britain

In Britain, 'creationsim' has been a much less influential movement than in America. Evolution was not so much argued about as taken for granted, and it was taught without dissent in biology classes. An Evolution Protest Movement (EPM)[8] was founded in 1932 with the stated aims of publishing scientific information supporting the Bible and demonstrating that the theory of evolution was not in accordance with scientific fact. But, although it numbered some eminent scientists among its members, most of the literature of the EPM was written by non-scientists, and the criteria for assessing truth tended to be Scripture rather than nature.

Although creationist teaching has never been accepted by more than relatively small groups, the reverberations of the campaign rumble on. The confrontation between Thomas Henry Huxley and Bishop Samuel Wilberforce at Oxford in 1860 is repeatedly recalled by the media, although the fact that it was preceded by an hour's monologue by the arch-confrontationalist J. W. Draper (p. 256) is not. Moreover, it is wrong to regard the debate as primarily about science and religion. Wilberforce was concerned with the apparent legitimization of change produced by evolutionary ideas, at a time when church attendance was falling as men and women flocked into towns and the Bible's authority was challenged by 'higher criticism'. In contrast, Huxley's target was the English ecclesiastical establishment. For him it was not so much a conflict between science and religion as a campaign *in defence of scientists*, ' a sustainable battle for the cultural supremacy of the scientific community in late Victorian England'. Huxley had the laudable aim of achieving the maximum enjoyment of science by as many people as possible, but to achieve this he believed he had to confront religious dogma and proclaim his alternative gospel of 'scientific naturalism', conceived as a worldview that acknowledged the methods of science as supreme and which repudiated the supernatural. Huxley's 'Enemy' was the English ecclesiastical establishment and (with Draper) included the Church of Rome. There were, however, other enemies within the citadel of science regarded by Huxley as hopelessly reactionary, notably some of the leading scientists of the day working at Cambridge: George Stokes, P. C. Tait, William Thomson (Lord Kelvin) and James Clark Maxwell. In confronting these Christians, Huxley allied himself with like-minded thinkers in the high-profile X-Club, which consisted of nine distinguished men of science, all but one of whom were Fellows of the Royal Society and all of whom shared a commitment to scientific naturalism.

Huxley had no qualms about self-consciously taking every opportunity to invade what he saw as the territory of religion in order to expand what he spoke of as 'the church scientific', within which he happily referred to himself as a 'bishop' (L. Huxley 1908: 251–252). In like manner, Francis Galton spoke of a 'scientific priesthood' (Galton 1874: 260). For Huxley and those like him, 'there is but one kind of knowledge and but one method of acquiring it' (T. H. Huxley 1870: 22), and their science was both empirical and reductionist. On this view, even human beings could be reduced first to animals and ultimately to machines.

But they were men of the time. Russell notes that 'to modern historians of science, well aware of their own fallibilities, Huxley and his X-Club members had no monopoly of the desire for a free science' (C. A. Russell 1985). More than two centuries earlier, as Hooykaas (1957: 20) has reminded us, the Puritan scientist Nathaniel Carpenter had written: 'I am free, I am bound to nobody's word, except to those inspired by God; if I oppose these in the least degree, I beseech God to forgive me my audacity of judgment, as I have been moved not so much by longing for some opinion of my own as by love for the freedom of Science.'[9] And, as Russell has written of Joseph White's *History of the Warfare of Science and Theology*, 'the factual errors (and there are many) are less reprehensible than the approach to history that informs the whole of White's book. In a word, this is a Whiggish historiography, seeing only progress in science, and impervious to the possibility that theological arguments might have any real substance, let alone that science might in any sense be indebted to theology' (C. A. Russell 1989: 24). Despite the still all too popular conflict metaphor beloved of the media, we nevertheless believe that a biblically based theology is not only plausible, but, on the evidence, remains a key feature in the development of science.

According to John Hedley Brooke, three distinctive themes recur in the historical relationship between science and religion. The first is the notion that there is an inevitable conflict between the two; but this view has been undermined by historians of science over the past few decades. The second is that if only scientists and theologians had formulated their statements more clearly, they would have realized that they were complementary, not conflicting. This oversimplifies the issues, since ideas have steadily passed between science and theology over the centuries. The third is that religious beliefs and scientific claims have in fact often affirmed one another. In Brooke's overall assessment, 'Serious scholarship in the history of science has revealed so extraordinarily rich and complex a relationship between science and religion in the past that general theses are difficult to sustain. The real lesson turns out to be the complexity' (Brooke 1991: 5). This is a timely reminder of the ever-present danger of seeking to use the history of science selectively so that it is 'hijacked for apologetic purposes' (Brooke 1991: 42). Notwithstanding, we must not diminish the validity of the generalization that the birth of science in the seventeenth century was significantly and profoundly influenced by theological concerns. This is a truism to historians of science, but may yet come as a surprise to scientists indoctrinated by the conflict metaphor and still believing that science offers explanations competing with those of religion.

Chapter 2

God, creation and the laws of nature

We all devise ways, however rough and ready, of organizing and storing our moment-by-moment experiences. Rules of thumb about seeming regularities may become increasingly sophisticated, from the simple and concrete pictures of children and primitive peoples to the highly abstract and complex models of the mathematical physicist. The Bible offers a gallery of portraits and pictures which help us to understand the relation of God to his creation. They help us to know how to think about miracles and their relation to the so-called laws of nature. How do the biblical authors refer to and use the occurrence of miracles? These are some of the issues discussed in this chapter.

In order to make sense of the range and variety of experience which we gain daily, we all employ some form or other of organizing and storing such experience. These shorthand accounts can be described as thought-models; in practice we use them most of the time. Some become highly organized, systematized and quite explicit; others remain vague and usually implicit. Their function is to produce a consistent structure to co-ordinate and make sense of our knowledge and experience. They vary enormously, of course, in their complexity. Children and primitive peoples tend to use very simple and concrete pictures. At the other end of the continuum we find mathematical physicists employing highly abstract and complex models.

What is true of our experience in general is also true of our experience and knowledge of the material order in particular, and, when we stop to think about it, it is true also of our thinking about God and his relation to his creation. As we saw in the last chapter, a wide range of different thought-models have been used down the centuries concerning the relation of God to his creation. In a sense the emergence of each new model served to underline the fact that any model, however refined, is never fully adequate for the task. And arising out of these models has been a range of views concerning the meaning and the status of the so-called laws of nature. In recent years historians of science have shown how some of our earlier tidy analyses failed to reflect what was actually behind debates about the laws of nature.

Gaining historical perspective

People, whether scientists or not, may mean very different things when they speak about the laws of nature. There is an ever-present risk of turning *laws* of nature into laws of *nature*. Such a change implies that a sovereign creator is bound by his laws and at the same time relegated to the periphery as merely a 'first cause'.

This is well illustrated by Darwin, who wrote in his *Autobiography* that 'the more we know of the fixed laws of nature the more incredible do miracles become'. John Hedley Brooke traces these views back through thinkers of the eighteenth century, who regarded natural laws as little more than a code for affirming the continuity of nature, but points out that there was no consensus among seventeenth-century natural philosophers about the significance of the laws of nature.

> They could be understood, as they were by Galileo, as mathematically expressed idealizations against which the real world could be compared. They could be understood as divine commands, as they were by Descartes; or merely as the 'rules' by which nature operated according to the divine will. This last, more modest formulation, was that of Robert Boyle, who was worried that matter was too stupid to know what a law was! (Brooke 1991: 8).

Even worse, a double meaning may be attached to the 'laws of nature'. William Whewell, Darwin's respected Cambridge contemporary who was a philosopher and Anglican clergyman, wrote: 'With regard to the material world, we can at least go so far as this – we can perceive that events are brought about not by insulated interpositions of Divine power, exerted in each particular, but by the establishment of general laws.' Darwin used (or perhaps we should say misused) this quotation by placing it opposite the title page of his *Origin of Species*. Darwin wanted to explain the origin of species *without* reference to divine intervention; Whewell, in his Bridgewater Treatise (*Astronomy and General Physics considered with Reference to Natural Theology*, 1833), strove to maintain the limits of law-based explanations in the biological sciences by insisting that the origins of living things remained a mystery. So Darwin in public was concealing something that was well known to Darwin in private. There was a duality, an ambivalence within Darwin himself. He would refer, as he did in the *Origin of Species*, to 'laws impressed in matter by the Creator', and then have second thoughts – not about the impression by a high power, but about his use of the 'Creator', which might mislead his readers into believing that he favoured a Christian orthodoxy. While he denounced the enterprise of natural theology as vain and anthropocentric, he seems to have retained an unspoken conviction that, in Gillespie's words, 'rationality and moral probity of God underlay the rationality and meaningfulness of science' (Brooke 1991: 86, 90). It is perhaps not without significance that *the* major figure in biology never seems to have had a coherent philosophy of nature, let alone a theology of nature.

The implication is that the erosion of religious belief by science is a view encouraged by the assumption that we are continually discovering more laws of nature. Brooke concludes:

> The assertion that science has threatened religious values by subsuming natural phenomena under physical 'laws' is seductive and appropriate in some cases. But it is at best a partial account of a complex historical process in which debates about the meaning of the word 'law' were crucial. One could use the word to stress not the autonomy of nature but its dependence on a divine legislator. As with Whewell in the first half of the 19th century, the concept often had explicit theological connotations and helped to excite not suppress a sense of awe. Even in Darwin's mind, the existence of laws governing an evolutionary process did not necessarily rule out higher ends (1991: 102).

Brooke believes that we find the real issues separating secular and sacred readings of nature behind what he calls the veil of natural laws, and these issues have to do with 'debates about the ultimate nature of the agencies and powers, the regularity of whose manifestations is summarized in the articulation of physical laws' (1991: 102). If he is right, there was not, as is often asserted, a steadily advancing tide of scientific law.

Modelling the laws of nature
Of the various models lying behind the notion of laws of nature, two have received special attention in the past. Taken to extremes, they imply theoretically indefensible views of the relation of God to his creation.

We saw in chapter 1 how the picture of God as the 'divine mechanic' or 'machine-tender' emerged in early modern times (p. 23). The main properties of this sort of model, which we can call the *craftsman model*, are derived from our human experience of what it means to create something out of existing materials, whether it be an instrument or a complicated machine; God the Creator is conceived as an infinitely wise and clever inventor and constructor, who has produced the Universe as we know it and has set it running. It is usually also assumed that once the machine has been set running, it is to all intents and purposes autonomous, except for very occasional *interventions* when some particular event has to be brought about or some servicing of the machine is required. Hooykaas (1971) has helpfully distinguished deism (referring to the view that God once and for all creates the Universe, and then leaves all to the Universe he has 'programmed') from semi-deism (which refers to the situation where God created once and *occasionally* intervenes).

When the 'divine mechanic' model is applied in the theological context, we find that it can lead to undue emphasis on two aspects of God's relation to his creation. In the first place, it tends to over-emphasize God's aloofness from that which he has created; and secondly, it implies that it is necessary for God to intervene from time to time in order to bring about occasional unexpected occurrences for his particular purposes, so that it is only in these occasions that we clearly see his divine activity. The most celebrated and widely known of the models of this kind was that proposed by William Paley, who compared God to a watchmaker and that which he created to a watch. Such a view leads to a false kind of supernaturalism, in that it suggests that God's activity in the created world should be looked for mainly in occasional acts from without, injected into the otherwise autonomous orderly working of the machine. In short, models like this encourage a philosophy of nature which regards it simply as a machine which needs no divine sustaining activity to keep it in existence from moment to moment. So for God to bring about events which are commonly considered to be *miraculous*, he must return and intervene in a system which he has previously set going and then left. This is the view Hooykaas (1971) refers to as semi-deism. Deists, with their commitment to uniformity as an expression of the divine power and wisdom, tend to argue that miracles are ruled out on *a priori* grounds, in that were they to occur, they would show that God needed to have 'second thoughts' about some aspect or other of his creation – and that would reflect adversely on him. It is also a largely unitarian model, ignoring the redeeming and sustaining activities of God.

Another feature of such models is that the craftsman or mechanic who makes artifacts makes them primarily for a particular use. They are not an end in themselves, and consequently express little of the personality of the craftsman. Such models therefore encourage one to ignore or at least minimize questions concerning the purpose for which the created order exists.

There are, however, certain aspects of this kind of model which do focus our attention on important theological truths. It reminds us that the purpose of any instrument or machine lies outside the actual constitution and mechanism of the machine itself; the purpose in fact lies solely in the mind of the creator and not within that which is created. This highlights the truth that the creation is instrumental to God's ultimate purposes, designed to achieve an end beyond itself. The Westminster Confession summarizes such a purpose powerfully: God created the world 'for the manifestation of the glory of his eternal power, wisdom and goodness'.

A second group of models concentrates on God as a creative artist, and as such conjures up for us a different set of ideas. Models of this kind were championed by Dorothy L. Sayers (1946), and, in more recent years, have been developed by writers such as Donald MacKay. They differ in many ways from the craftsman model. For example, a work of art has worth within itself, rather than for any instrumental use to which it is put; it exists for enjoyment and admiration, more than for use. Moreover, its inherent value lies in what it reveals of the truth which the artist wants to express, and also (and importantly) in what it shows of the character of the artist himself. Applying this to God, we recognize that, like the artist, God does not *need* to create in order to understand more clearly for himself the truth which he wishes to express; he creates from pure love of creation. Notwithstanding, there is a sense in which God also may contemplate his work and say that it is 'very good' (Gn. 1:31). Indeed, the distinction between revealing some aspect of the truth and expressing the character of the artist disappears when we think of God's creative activity. In his creation God certainly reveals something of his majesty, power and Godhead (Rom. 1:18–23), but this revelation also expresses his character. Models of this kind clearly lay much greater emphasis upon the immanence of God than does the craftsman type of model, which emphasizes his transcendence. They carry the danger that, when extended beyond their primary purpose, they can all too easily slip into pantheism.

In order to do full justice to our understanding of the relation of God to his creation, we have to use a variety of models, each focusing on a particular aspect of the relationship. For example, both the 'craftsman' and the 'creative artist' models share a radical shortcoming in failing to do justice to the clear biblical teaching that God continues to sustain the Universe and to hold it in being moment by moment; they leave us with a picture of the creator completing his instrument or his work of art and then leaving it at that. How may we come a little nearer to doing justice to the continuing activity of God in relation to his creation? It is at this point that an elaboration of

Dorothy L. Sayers' approach by Donald MacKay is helpful, albeit in some aspects controversial.

MacKay (in Jeeves 1969: 23) invited his readers to adapt their thinking about creative artistry to modern technological developments of the mid-twentieth century. This takes up aspects of another analogy, that of a musician performing; the sound continues only so long as the musician continues playing. We can extend this: imagine that instead of our artist using canvas and oils, he uses a television screen to display his creation and he uses the transmitting apparatus of the television station in order to generate the display which he wishes to portray to us. The important difference between this variant of the 'creative artist' model and that of the more traditional artist is that the picture on the screen continues to exist and to have its present form only as long as our artist continues to generate the programme which expresses his mind. The moment he stops, our picture ceases to exist. This model thus solves the problem in our earlier 'creative artist' model, in that it underscores the continuing activity of the artist in holding his creation in being from moment to moment.

Even so, this model of the creative artist still has a major inadequacy, which MacKay addressed to make it closer to the biblical picture of God's relation to his creation.

We must first of all recall how the Bible opens with a narrative about the creative activity which gives rise to the existence of our world and its inhabitants. The Bible reminds us frequently (*e.g.* Col. 1:17; Heb. 1:3) that not only do the objects of creation, including ourselves, owe their continuing existence to the activity of the creative Word, but also the whole space-time meshwork of events are 'upheld' and 'cohere' by and in the same creative Word. More important still, the biblical picture teaches that our Creator is active within the drama of our existence not only in his creative sustaining power moment by moment, but also in some mysterious way in his personal self-revelation. As MacKay wrote:

> Our Creator is more than simply the Creator of our drama, he is also the Creator-participant. With this in mind we must also note that nothing we say on the one hand about our createdness, must be allowed to distort or diminish the truth which is conveyed to us on the other hand, in the many complementary pictures, which depict us as children of a loving father, as sheep that have gone astray, as prodigal sons offered a loving welcome in the home of our father (cited by Jeeves 1969: 24).

With this in mind, MacKay developed his earlier analogy of the creative artist: 'Let us imagine the relationship of the author, as a creator, to the literary work which he creates. We can notice certain relevant features of this at once, such as that our author, when he eventually conceives and utters his literary work, does so as a single coherent picture, including the past, present and future of the characters of his story, and the world in which he sets them. This fact helps us to appreciate the logical distinction between the creator of the drama, who is in this sense a spectator, and that of the actor within the drama. We shall return to the relation between creator-talk and creature-talk, but for the moment we simply wish to note that it is a real distinction.

The next step is to imagine a character in our literary work who finds himself addressed by his fellow characters, some of whom claim to speak to him in the name of his creator and their creator. This refers, of course, to the way in which from time to time the prophets spoke and prefixed their statements with words such as 'Thus says the LORD . . .' 'Most amazing of all, the character in our literary work suddenly finds himself confronted by, and personally addressed by, a fellow character who claims to be identical with the creator of the whole literary work and all its characters. Here we are already involved in the mystery of the Incarnation.' In this way MacKay steadily adapted and improved the picture of the creative literary artist, in order to do justice to the biblical teaching which declares that God in eternity, our creator, is also identical with the one 'who spoke by the prophets', who was in Christ reconciling the world to himself, and who still today continues to invite personal dialogue and personal relationships with the creatures he has made.

We shall have occasion to refer back to this creator-participant model of MacKay's in later sections of this book, so it is relevant to identify important features of it, as an aid to disentangling (or at least setting into a new perspective) some of the recurring debates concerning topics like free will and determinism, providence, and the laws of nature. Crucially, this picture alerts us to distinguish carefully at all times the different logical standpoints of the creator and his creatures. This is especially important whenever we are tempted improperly to oppose creator-talk and creature-talk.

For the divine Creator, his work is a space-time unity; it is created as a whole. He creates the space-time framework of his production, and not only does he create the scenery, the setting, and the characters of his work, but also, and more significantly, he creates the events that occur. The creator does not merely leave the characters, the scenery and the environment of his work to interact according to some 'natural law' apart from his control, but, much more significantly, he creates the events past, present and future, and holds the whole thing in being from moment to moment in this sense. He creates the timescale in which the events of his characters take place, but also undertakes his own creative activity *outside* this creaturely timescale. We shall not explore the details of this model further now, but shall see how the

Christian model of the relation to God to his creation encourages us to envisage the so-called laws of nature, and the implications of this for questions of miracles.

Miracles and the laws of nature

It is often asserted that there are too many problems for reasonable people in believing both in the lawfulness of the natural order and in miracles. A number of questions always seem to be raised about whether belief in the lawfulness of nature *leaves room* for God to *intervene* in the natural order; or whether God *uses* natural laws in order to bring about his creative purposes; or whether we should regard miracles as God's interventions in the otherwise orderly working of creation. Many will recognize ways in which these questions are solemnly asked and often firmly answered one way or the other. Yet if the sort of approach we have outlined above approximates to the truth, then, as MacKay emphasized, ideas such as *leaving room* for God, or God *using* natural laws, or God *intervening*, must all stem from a model or models which are intrinsically inadequate. Indeed, they may not merely be inadequate, but also frankly misleading.

An essential point in any model of God's relation to his creation is that according to the Bible, nothing continues to exist or continues in being apart from God's moment-by-moment activity. It therefore becomes meaningless to ask whether the laws of nature *leave room* for God's involvement. How could they *leave room* for God's activity, since God's activity is present all the time? Or again, how could God *intervene* and *suspend* his laws from time to time, since he is there all the time, holding everything in existence? In what sense could God *use* natural laws, since natural laws are only our way of summarizing our experience of the regular occurrence of events in the creation which God holds in being all the time? The expressions *leaving room*, *intervening* and *using* assume and condone a radical misconception of God and of his relation as creator to the created order. We believe that the biblical view requires that the whole pattern of space-time events is not only conceived but also held in being moment by moment by God; it is incorrect to term miraculous events as interventions. They are in fact no more and no less dependent upon God's activity than day-to-day occurrences which we so readily take for granted, like boiling a kettle or riding a bicycle, even if we use shorthand language to summarize our experience in terms of what we call natural laws.

Let us go back to the model of the electronic artist introduced earlier in this chapter. Consider the situation in which we may be watching a television production of a new play. Let us also suppose that the play is written and produced by the owner-operator of the television transmitting network. As we watch this production, we are shown a long sequence of a hitherto unknown game. Intrigued by this game, and interested in playing it ourselves, we watch it very carefully and make detailed notes of what happens. Eventually we believe that we have deduced the rules of the game from observing carefully the regularities in the events which have occurred during it, and that we can to some extent predict the reactions and initiatives of the players. Our position as viewers is in some important respects similar to our activity as scientists, observing the events in God's creation and then attempting to discover the rules of how these events hang together. Of course, in other respects this picture is inadequate, since as scientists we not only observe but also manipulate the events, and this we cannot do in our analogy.

In a very limited sense, the owner-operator-author and producer of the production that we have been observing on our television screen has similarities with God's creative and sustaining activity of the events in the world around us, since if the transmitting station were to go off air, there would be nothing left for us to observe. The show would be over. Regarded in this way, our discovery of the regularities in the events we have been observing means that, for us, natural laws are *descriptive* rather than *prescriptive*, and emerge for us only *post hoc*, as features of and within the created order. This helps us to understand how miracles relate to natural laws. It should be clear that it is not only improper, but also mistaken, to focus attention on miracles entirely or even principally on the grounds that they give *factual* evidence for divine activity. If we give full weight to the biblical assertions about God's divine upholding of all things at all times, then the events that we label miraculous, as well as routine, non-miraculous events, are equally dependent upon the same creative power of God. We do better to focus on the mode or purpose of God's activity in those events that we consider to be miraculous than on the lack of correspondence to known scientific laws of events that we label miraculous. This means that we need to clarify our use of the term 'miracle' before discussing miraculous events further.

Since all events are dependent upon God's continuing activity, the term 'miracle' is best reserved for those events which reverse our normal observations or expectations, retaining 'providence' to cover the daily gifts which we constantly receive at God's hands. How are miracles portrayed to us in the biblical record? The first thing to notice is that the Bible does not focus upon the relation of the event we call a miracle to the natural order, but rather upon the impression which that event made upon the minds of those who witnessed it (*e.g.* Mk. 3:11; Ex. 14:31). We find also that the relation which the miraculous events bear to the wider purposes of God's revelation of his will is emphasized. The biblical miracles therefore direct attention to the impression that the event makes upon those who witness it, rather than to theoretical questions, such as whether the cause of a miracle is regular but still unknown to us, or whether it is in some way contrary to our normal expectations.

An examination of the Bible soon convinces us that to label an event as miraculous does not imply that there are no known natural causes for that event. On some occasions we are invited to notice the natural cause which was responsible for the event. The crossing of the Red Sea is especially instructive in this regard, since the cause of the rolling back of the waters is stated in Exodus 14:21 to be a strong east wind. This aspect of the occurrence of miracles has encouraged some people to regard such miracles as nothing more than divine coincidences. It does, however, remain the case that it is only against the backdrop of what we have already come to expect of the regular workings of creation that we can perceive the unusual events to which we attach particular significance and regard as miracles.

Miracles are certainly not invasions by God into an otherwise natural working of creation, for this would deny that in some sense God is there already. Neither are they merely natural, if by this we imply that God is not active in the whole stream of events moment by moment. They are, rather, special acts of God, and seen to be such – albeit acts in which the secondary means which are responsible for the event are neither more nor less given by God than any other day-to-day occurrence.

It may perhaps be helpful to note that there are some miracles which have some unusual awe-inspiring and distinctive feature, which results in their being regarded as omens or portents of something yet to occur (for instance, Elijah on Mount Carmel). Nevertheless, their primary purpose is to fix attention upon the message that accompanies the event. Such events are open for all to see, as distinct from events which are miraculous only through the eyes of faith. It is clear from the Old Testament record that the miraculous preservation of Israel over the centuries excited the admiration of the people of surrounding unbelieving nations, as is made clear in Joshua 2:10.

A second recurring feature of some miracles is the way in which they are seen as mighty acts of divine power. As Clark Pinnock has written: 'The mighty acts of Jesus were performed by One who is Himself called the power of God, and these works were entirely appropriate actions to be performed by One who was both Man and God. As some have put it, they served for him as credentials in the midst of an unbelieving generation' (quoted in Jeeves 1969: 30).

Finally, an all-pervading characteristic of miracles is their importance as signs, tokens or pledges of an age yet to come; this is true in both the Old and New Testaments. Indeed, some would regard this as the key aspect of miracles. Thus the healing miracles are seen as a temporary rolling back of the claims of death, which will one day be abolished.

All three of these features of miracles are present in the healing recorded in the third chapter of the Acts of the Apostles. We are told that those who observed the miracle were filled with wonder and amazement; the apostles made it clear that this mighty act of divine power was, as they put it, 'not by our own power', but by the power of Christ; and finally we may see this as a sign and pledge of an age yet to come, when all disease and sickness will be done away with.

It is sometimes not realized how relatively scarce the miracles are within the biblical narrative as a whole. Put another way, if we were today writing a narrative with the express intention of impressing upon our readers the other-worldliness of the events that were portrayed and the claims that were made for what was said, we should be sorely tempted to ensure that our narrative was well stocked with miraculous events. When we consider the thousands of years covered by the biblical narrative, we find that this is not the case. The miraculous events tend to concentrate around three major eras in the total biblical record, namely the events of the exodus of Israel from Egypt, the time of the prophets of the ninth century BC, and the apostolic era recorded in the New Testament. Such outstanding biblical characters as Jeremiah and David, for example, have no miraculous acts attributed to them.

The biblical records do not suggest that miracles should be used as knock-down arguments or incontrovertible evidence, although on one occasion they are described as 'infallible proofs' (Acts 1:3, AV). It is almost certainly the case, however, that we too readily give the word *proof* a juridical or even a scientific meaning. This is familiar to us today, but would have been quite foreign to the minds of first-century readers. If this is so, then the function of miracles should be regarded not as providing incontrovertible proofs, but as events which bear witness to the divine character of something that is being proclaimed, and/or of the person who is proclaiming the message. Another aspect of this is that there may be differences of interpretation about any factual event. As Aristotle put it, 'It is not the facts which divide men but the interpretation of the facts.' If a confirmed atheist had been present at a resurrection appearance of Christ, he might, on the basis of his own presuppositions, quite reasonably have exclaimed simply, 'I always thought there might be ghosts, and now I am convinced.' The point is that he would not inevitably interpret such a resurrection appearance as evidence for the divinity of Christ, but merely for the existence of ghosts. This, of course, serves to draw attention to the fact that many people turn away from the idea of miracles, not because they are scientists, but because they begin from an atheistic worldview.

It is ironic that the Christian view of miracles is more open-minded than the non-theistic point of view. While observation can tell us what has happened, it cannot tell what could have happened. This is the reverse of the normal assumption, and it is worth exploring. The most common view of miracles, especially as regards their apologetic value, derives from the philosopher David Hume. Writing in the mid-eighteenth century against a prevailing tradition that miracles as recorded in Scripture afford proof for the Christian faith, Hume asserted that 'We may establish it as a maxim, that no human testimony can have such a force as to prove a miracle, and to make it a just foundation for any such system of religion' (1748/1975: 127). In other words, it is a circular argument to seek proof of the truth of the Christian revelation by drawing from the pages of that supposed revelation, since it appeals to miracles contained in that revelation to authenticate the said revelation. Hume's logic fails, however, if the historical trustworthiness of Scripture could be persuasively argued on other grounds. Hume is concerned to undermine the reliability of such testimony, although he concedes that a strong consensus in the reports of a rare event or miracle might persuade us that it did in fact occur. Paul Helm comments:

> What Hume attempts to do is undermine the historical trustworth- iness of any testimony, including of course written testimony, which includes an account of that happening of miracles. If Hume's arguments are sound here, it becomes impossible first to treat the Bible or any other document as historically reliable and then to consider further evidence from miracles as giving authenticity to the documents as a divine revelation. The very fact that they contain miracle stories, and present these not as myths but as historical occurrences, debars the documents from serious consideration as historically trustworthy, and hence presents the argument from their trustworthiness, via the miracle stories, to their position as a divine revelation . . . [But] The Christian faith as it is derived from Scripture is inherently or essentially miraculous. For at the heart of the biblical account of human redemption are miracles; the miracle of the Incarnation and of the Resurrection of Jesus Christ from the dead. These miracles are not a prelude to anything else, they are the warp and woof of the Christian faith . . . When Nicodemus said to Christ, 'We know that thou art a teacher come from God: for no man can do these miracles that thou doest, except God be with him' (John 3:2), was he not arguing along similar lines? And are we not invited by the gospels to endorse this argument?
>
> There is a world of difference, however, as even Hume would allow, between a person who witnesses a miracle, and someone else who believes a miracle because of the testimony of someone else. The position of Nicodemus and of ourselves is therefore crucially different.

Furthermore, there is no problem of circularity with Nicodemus' claim (Helm 1987).

Hume certainly considered that the main thrust of his essay, and what he regarded as the originality in his contribution, was the probability argument – that when a miracle is reported it is *more probable* that the witnesses were deceived in some way than that the event occurred as they report it.

Despite the circularity of Hume's argument against miracles, it is still commonly used today, and many Christians remain confused about how properly to regard the miraculous. We need to be clear: rather than regarding miracles as incontrovertible arguments which warrant special treatment, they should be seen as an integral part of a coherent whole. Certainly Jesus did not pursue a policy of maximizing miracles. For example, we read that 'some of the Pharisees and teachers of the law said to him, "Teacher, we want to see a miraculous sign from you." He answered: "A wicked and adulterous generation asks for a miraculous sign! But none will be given it except the sign of the prophet Jonah" ' (Mt. 12:38–39). In Luke 16:31 we read, 'If they do not listen to Moses and the Prophets, they will not be convinced even if someone rises from the dead.'

Miracles in Scripture show a continuum of events which are at odds with scientific explanations to varying degrees. Scripture seems to regard it as unimportant to draw hard and fast lines between 'miracles' and other events. Indeed, miracles quickly become misunderstood and misused when they are separated from the rest of the biblical story; it is the coherence of the whole narrative which carries with it a convincing and persuading power. The 'natural' events of Christ's crucifixion and the 'supernatural' one of his resurrection make complete sense when understood against Christ's own teaching about these events. For example, in Luke 24:26 he asks, 'Did not the Christ have to suffer these things and then enter his glory?' Peter too sets these events in their overall salvation context: 'But God raised him from the dead, freeing him from the agony of death, because it was impossible for death to keep its hold on him' (Acts 2:24).

Miraculous events in the Bible also resonate with the main thrust of the character of the God pictured in Scripture; the emphasis throughout is of God graciously bestowing blessing upon undeserved blessing upon humankind. It is wholly mistaken to separate miraculous events from the overall activity of a gracious God and his teaching.

It is worth noting that there is a category of 'miracles' used frequently in contemporary discussions that is of our own making and not one encouraged within Scripture. John Polkinghorne reminds us:

> The history of science is full of the unprecedented and unexpected.
> Electrical conduction in metals behaves in an orderly way until

suddenly, below a critical temperature, some metals lose their electrical resistance altogether and become superconducting. Changing circumstances (in this case, lowering the temperature) have created a new regime in which physical behaviour is suddenly different. As the physicists say, a phase change has occurred. The task of the scientist faced with such a phenomenon is to try to find the underlying regularity which embraces both the old familiar regime and the strange new one revealed to him. The coherent achievement of that task is what constitutes the advance of scientific understanding. Ohm's law is not an unbreakable law of the Medes and Persians, but it can be subsumed into a wider framework of order. In the case of superconductivity it took more than fifty years from the discovery of the phenomenon by Kammerlingh Onnes to its explanation by Bardeen, Cooper and Schrieffer (Polkinghorne 1983: 55).[1]

In the ministry of Christ, miracles provide, in Polkinghorne's words, a coherence in understanding that ministry, in which 'even the winds and waves obey him'.

This brings us back again to the laws of nature. It is often claimed that because of the laws of nature a particular miraculous event could not have happened. This use of the term 'a law of nature' differs from that of the scientist, to whom it is a regularity of a statistical kind; it is not something that is logically necessary but one which is contingently true at a basic level of generality. When Hume referred to 'violating' a law of nature, he seems to understand a law of nature as being like a law of Parliament. If Parliament decrees that you shall drive on the left-hand side of the road and you drive on the right, then you are indeed violating the law of Parliament. For some people it seems that what Parliament says about which side of the road you drive on is analogous to what God says about the laws of nature. But that is a misunderstanding – a law of science is not like a law of the land. A scientific law, while stronger than a mere regularity, is weaker than a law of logic; it is *descriptive* rather than *prescriptive*.

A 'law of Parliament' concept of a law of nature tends to suggest that the uniformity of nature ought to be defined in such a way as to exclude the possibility of miracles. In contrast, the Christian viewpoint is less restrictive; it agrees that it is perfectly legitimate to assume uniformity in nature, but is willing to entertain the possibility of non-uniformity (or miracle), if there are good grounds for doing so. In other words, our conception of natural laws acknowledges that they are based on a finite number of observations or experiments, and that they must always remain subservient to, rather than normative over, any further observation. Hence theists are more open-minded towards historical events than non-theists committed to uniformity, who must do their utmost to explain away happenings which do not fit with their presuppositions. They have to do this because they are already committed to a metaphysical principle of uniformity. (There may, of course, be atheists who are equally committed to the place of chance and the inexplicable in nature.)

In conclusion, it may be helpful to set out some implications for the way miracles may be presented in the proclamation of the Christian message. From the human standpoint, the progression from unbelief to belief may involve (in the first place) only a minimum of information about the Christian gospel; in the light of this minimum of information it seems reasonable and fair-minded at least to consider the possibility that the message of the gospel may be true. It is frequently the case that the testimony of others who have walked this way before and have become Christians, and who now believe in miracles, will help the enquirer at least to consider the reasonableness of the theistic point of view. By implication, this carries with it the realization that if the Christian view of God and his relationship to his creation is correct, then God not only can, but may, do miracles from time to time. These unusual events will inevitably appear contrary to our normal, day-to-day experience. When people become Christians and read and study the biblical record in the new light given to them through their relationship with Christ, miracles begin to make sense as part of a coherent whole. They become all of a piece with what one 'knows' about the gracious moment-by-moment caring and upholding of a loving Father God, who was in Christ, walking, healing and teaching in Galilee two thousand years ago.

Chapter 3

The scientific enterprise

What do we mean by the scientific enterprise? Is there such a thing as the scientific method? How do the contributions of the historians and philosophers of science help in formulating a view of the scientific enterprise acceptable to practising research scientists? What are meant by the labels 'worldviews' and 'world pictures'? How does the scientific path to knowledge relate to other paths? Where do values enter into the scientific enterprise? This chapter argues that only by answering these questions can we avoid making excessive and unjustified claims about the nature of the scientific enterprise and of the knowledge to which it gives rise.

No-one today can fail to be impressed by the remarkable achievements of scientists. They have been responsible, in one way or another, for the cars we drive, the videos we use, the jet planes we fly in, the medical discoveries that alleviate our sicknesses and the methods of education which facilitate learning at all ages. Such past achievements generate expectations of future benefits. It is natural for us to have a high view of science as a source of secure and useful knowledge. In contrast, the impact and benefits of the arts, the humanities and religion seem, at times, to be meagre and ephemeral. Science has come to be seen as a sacred cow which, if appropriately fed, will continue to yield tangible benefits for the good of all. Add to this the fact that in the teaching of science we ordinarily limit our laboratory exercises to straightforward and relatively simple experiments which produce easily gathered, clear-cut results, and it becomes almost inevitable that science should be viewed as the mechanical application of a foolproof method; by its very nature it will continue to give valuable and unchallengeable results. Hence the widespread conviction that scientific knowledge is more reliable, more trustworthy and more beneficial than any other kinds of knowledge available.

This view is reinforced by the way in which research scientists report their results in learned journals. In the interests of efficiency of communication, and mindful of the limitations of space, we leave out all the personal element. We set out our methods, our procedures, our results, our analyses and our conclusions in as succinct and unambiguous a way as possible. We omit the false starts, the ineffective methods we tried and then discarded, and all the intuition, imagination and casual discussions with colleagues at conferences and over coffee which affect every part of our work. This must be so in the interests of clarity and brevity. And yet any scientist actively engaged in research knows perfectly well that if the nature of science and the methods it employs depended solely on the formal reports published in scientific journals, we would finish up with a complete misrepresentation of what science is really all about and how progress occurs.

When writing of the relation between science and Christian faith, it is perhaps understandable that most commentators concentrate on the content of particular scientific disciplines and the implications that they might have for traditional Christian beliefs. There is, however, another very important dimension to the relation of the scientific enterprise to Christian faith, well illustrated by the writings of philosophers interested in the logic of scientific discovery and the status of scientific assertions. This is the main subject of this chapter.

During the twentieth century, three such influences made a major impact on the way that apologists and theologians have thought and written about the verifiability and reliability of assertions traditionally made in theology: logical positivism or verificationism; falsificationism; and what we may call Kuhnianism. Philosophies of science have given rise to the common idea that science has undermined theology not because of the content of particular sciences and the relation of these to traditional beliefs, but rather because in some profound sense science has provided us with a touchstone of what is acceptable knowledge and rational belief. On this basis, many people assume that science points to principles by which all sound knowledge must be tested and established; since traditional religious faith is apparently unable to meet these criteria, it cannot be more than mere speculation. Thus it is not that the content of science conflicts with the content of religious beliefs, but rather that scientific statements are well established and theological assertions are not. Since the latter lack rational support, they must either be revised significantly or ignored altogether. This implies that scientific claims have a justification which is entirely different from theological ones; the philosophy of science, so it is assumed, demonstrates that science is about proof and reasonableness producing solid and secure knowledge, whereas religion is about faith and dogmatism, or perhaps even gullibility and matters of opinion only.

Some scholars have been overly impressed by these arguments about science. They have come to believe that the power of scientific method has so reduced the credibility of theology that it is hardly worth bothering to affirm or refute its claims. Indeed, some theologians have been so taken with fashions in the philosophy of science that they have accepted them without scrutinizing them or identifying their intrinsic weaknesses – weaknesses, apparent to the scientific community, which have led to a rethinking and reformulation, or in some cases rejection, of all of them. The three 'isms' listed above all have reductionist implications for theology, but we believe that none of them gives an adequate account of the nature of science and of the knowledge to which it gives rise.

Is there such a thing as 'the scientific method'?

An over-formalized view of the nature of science and the scientific method is not confined to the popular media. It is at times presented in courses on the philosophy of science, and in discussions of the scientific method in textbooks. The reason for this is not difficult to find. Philosophers of science, anxious to tease out the underlying logic of the scientific method, understandably often tend towards an idealized model of what the scientist has been doing. For good reasons, they almost always use as primary source materials scientific papers and books rather than the working notes or memoirs of practising scientists; not surprisingly, this, at times, gives a highly sanitized account of what really happened.

If some philosophers of science are ignorant of the way that science really works, it is equally true that the vast majority of working scientists remain blissfully unaware of the many perceptive writings by philosophers of science on scientific method and the scientific enterprise. Nevertheless, they continue to produce excellent science despite being unaware of the latest fashions among philosophers of science.

In the context of this book, our focus is whether we can characterize scientific knowledge in such a way as to understand how it relates to other forms and sources of knowledge, in particular religious ones. More specifically, we need to know how the account of the world we give through the scientific method relates to other accounts, such as those in Scripture. Must we choose between religious and scientific interpretations of the world, recognizing as we do the relative reliability and objectivity of scientific knowledge as compared with other kinds of knowledge? Simply to pose the question in that way assumes that we all know and agree on what we mean by 'objectivity'. Following David Hume in the eighteenth century, a distinction was made between the public, rational world of what is the case, and the private, subjective understanding of what ought to be the case. To Hume, 'morality is more properly felt than judged of'. The public arena became identified with science, and opposed to personal beliefs and their associated morality; religion was not merely unscientific, but beyond the scope of reason and perhaps positively irrational. This distinction was a major premise of the Vienna Circle of logical positivists during the 1920s, and the 'scientific worldview' came to be synonymous with reason and objectivity (see below, p. 55).

Reaction against the dogmatic positivism of the Vienna Circle (asserting that the only true form of knowledge is the description of sensory phenomena) has led to questioning whether there is a world to be investigated which exists independently of human belief and language, and hence to an 'anti-realism' which sees truth and reality as a construct of language rather than as something to be discovered. This latter approach has spawned the 'postmodernity' debate, with its assumptions that all beliefs are relative. While not seeking to evade a debate which is important for Christians, since acceptance of a postmodernist position implies that the idea of God is superfluous and that there is no firm basis for morality or for regulating society (Allen 1989), as scientists we must reject any suggestion that there is no 'real' world which can be investigated. Science legitimately claims to put forward views that are not tied to a particular perspective; what holds true in Washington ought to do so also in Nairobi and Jakarta.

Richard Dawkins is unapologetically forthright: 'Show me a cultural relativist at thirty thousand feet and I'll show you a hypocrite. Airplanes built to scientific principles work. The stay aloft and they get you to a chosen destination. Airplanes built to tribal or mythological specifications, such as the dummy planes of the cargo cults in jungle clearings or the beeswaxed wings of Icarus, don't' (Dawkins 1995: 32).

Scientific 'objectivity' does not imply a monopoly of 'objective' truth. There is an important sense in which religion is as objective as science, since both purport to be about objective reality. To argue that science deals with objective facts whereas religion is concerned solely with subjective values is to fall into the Enlightenment assumptions of the positivists.

Notwithstanding, the popular view is that science rests solely on precise observations, so that a scientist deals with 'pure facts' which yield indubitable, objective and impersonal knowledge. It is too often forgotten that there are no such things as uninterpreted facts. You do not have to be a psychologist to know that all our experience is organized on the basis of pre-existing interests and presuppositions, and that we attend only to selected features of our environment and ignore the rest. This applies in science just as much as in other activities, and deliberately so, otherwise we would be overwhelmed with the data that confront us. We concentrate on certain variables, and as far as possible select these out of other factors which we choose to control. But already we have brought our judgments into the equation.

Those persons, whether scientists or not, who characterize the scientific method as the orderly succession of systematic observation, formulation of hypotheses, testing and exploring their implications in further experiments, retesting and so on, often put forward an extreme objectivist view of knowledge. Such a view opposes objectivism to relativism and subjectivism, claiming that knowledge is connected with objective reality and is not simply what seems true to us. It is frequently associated with the assertion that the only statements which have any meaning are those that can be empirically verified, a view that held sway in the middle of the twentieth century. It implies that there is no possibility of valid knowledge which is not scientific, and that by sharing verification among a large number of people, greater objectivity is achieved. The result is that the accumulated knowledge gains an ever less personal character. On this view there is a clear contrast between (on the one hand) scientific procedure and the knowledge gained by its use, and (on the other) the erroneous hit-or-miss quality of personal judgments; it is assumed that only the former can provide objective, factual and reliable knowledge.

The history of science and the philosophy of science

If we reflect upon some of the epoch-making discoveries and contributions in the seventeenth, eighteenth and nineteenth centuries, we ought to remember that they were made in total ignorance of the analysis of scientific method available to us today. Older scientists took the world as they found it, forming hypotheses on the bases of the regularities they noted, carrying out experiments where possible and appropriate, refining their hypotheses, rejecting those that were false and modifying those that had some plausibility, and thus slowly moving 'scientific knowledge' forward. Their scientific method was not a formal enterprise, but it worked. And as their hypotheses were improved and extended, some were eventually given the name of 'theory' and then even of 'law'. But in practice the twin elements of experiment and theory are not easily separated, nor can the logical steps between the two be easily distinguished.

There has been a reaction against the formalization of science in recent years. Some of the leading philosophers of science have themselves drawn attention to the way in which accounts of *the* scientific method seem wilfully to ignore what has actually happened in the history of science. For example,

Dudley Shapere writes: 'The logical empiricist tradition has tended to lose contact with science, and the discussions have often been accused of irrelevancy to real science.' After discussing at length the views of Kuhn and Feyerabend, he comments that the relativism they both demonstrate 'is not the result of an investigation of actual science and its history; rather, it is the purely logical consequence of a narrowed preconception about what meaning is' (Schapere 1981: 31).

Assessing Karl Popper's approach to the logic of scientific discovery, Hilary Putnam comments: 'Popper's doctrine gives a correct account of neither the nature of the scientific theory nor of the practice of the scientific community . . . failure to see the primacy of practice also leads Popper to the idea of a sharp "demarcation" between science, on the one hand, and political, philosophical, and ethical ideas, on the other. This "demarcation" is pernicious, in my view; fundamentally, it corresponds to Popper's separation of theory from practice, and his related separation of the critical tendency in science from the explanatory tendency in science' (Putnam 1981: 69).

Imre Lakatos has noted that the kind of theories put forward by the likes of Ayer, Popper, Kuhn and others 'provides a theoretical framework for the rational reconstruction of the history of science'. But this leads to confusion: 'the internal history of inductivists consists of alleged discoveries of hard facts and of so-called inductive generalizations. The internal history of conventionalists consists of factual discoveries through the erection of pigeon-hole systems and their replacement by allegedly simpler ones. The internal history of falsificationists dramatises bold conjectures, improvements which are said to be always content-increasing and, above all, triumphant "negative crucial experiments".' He concludes: 'The history of science is always richer than its rational reconstruction' (Lakatos 1970: 154).

Widely discussed accounts of the nature of science

The three major philosophical descriptions of scientific practice are logical positivism, falsificationism and Kuhnianism. We begin with logical positivism. Its starting-point is a process of linguistic analysis which seeks to clarify the meanings of statements and of questions, and demands to know the criteria and procedures of empirical verification. It is asserted that the truth or falsity of statements must in principle be establishable by observation or experiment. In its original form, logical positivism was a systematic attack on metaphysics, in that it demanded that there must be observations in order to confer meaning on statements. A thoroughgoing logical positivist position necessarily rejects metaphysics as nonsense, since every basic statement must be verifiable in isolation in order to be meaningful. In fact, this argument has largely been abandoned, as many clearly important statements in science are not individually verifiable; and it has proved impossible to formulate consistent criteria of verification which are neither too 'weak' (failing to reject metaphysics) nor too 'strong' (rejecting important statements in science or other knowledge). Very few of the original logical positivists are still around, and even Professor A. J. Ayer, one of its leading proponents in the middle of the twentieth century, could be said to have seen the light. When asked in a widely publicized television programme what was wrong with logical positivism, he replied, 'I suppose the major problem was that it was pretty well all false.'

Logical positivism is little heard of today, and lacks influence in most philosophy departments, although in theology departments it is still upheld by some, such as Don Cupitt, and in psychology departments it is supported where old-fashioned behaviourism is still adhered to. More significantly, its influence is deep and continuing in our culture as a whole. This is well illustrated by the common, allegedly knockdown argument that there is a difference between religious claims and scientific ones in that only the latter can be proved or verified. It is this distinction which has led to the positivist approach being sometimes referred to as the verificationism programme, and which leads to the assertion that at root, religious beliefs turn out to be meaningless. The importance of verification was emphasized in Ayer's book *Language, Truth and Logic*. Ayer argued that religion was unable to meet the verifiability criterion of meaningfulness: 'We say that a sentence is factually significant to any given person if, and only if, he knows how to verify the proposition which it purports to express – that is, if he knows what observations would lead him, under certain conditions, to accept the proposition as being true, or rejected as being false' (Ayer 1946). The crux is the definition of 'verifiable'. Originally such verification was regarded as coming from sense data of one sort or another. From that starting-point any kind of theoretical talk based on non-observables was seen to be meaningless; this specifically included talk about God. 'To say that "God exists" is to make a metaphysical utterance which cannot be either true or false. And by the same criterion, no sentence which purports to describe the nature of a transcendent God can possess any literal significance' (Ayer 1946: 152). In the 1940s and 50s, some theologians retreated to argue that religious language should be reconstrued as a language primarily concerned with expressing values. For example, R. B. Braithwaite readily conceded that the positivists were right in their basic approach, so that religious belief became an 'intention to behave in a certain way (a moral belief) together with the entertainment of certain stories associated with the intention in the mind of the believer' (Braithwaite 1953). Michael Banner comments: 'Of course Braithwaite is really old hat now – so too are John Robinson, Van Buren and other "death of God" types who engaged in post-positivist theology. We know that the positivist programme is flawed, and so ought to feel unmoved by the arguments of those who presuppose it.' He goes on: 'Yet how many theological arguments still trade on positivist premises of a related sort?' Answering his own question, he targets Don Cupitt's book *The Sea of Faith* (1984), which criticized traditional faith in a way which 'only seems to make sense if there are positivist assumptions at large' (Banner 1990).[1]

Sir Karl Popper (whose views we criticize below) is unequivocal about what he calls 'operationalism or positivism'; according to Schapere, it was 'a fashion which Einstein later rejected, although he himself was responsible for it, owing to what he had written about the operational definition of simultaneity. Although, as Einstein later realised, operationalism is logically an untenable doctrine, it has been very influential ever since in physics and especially in behaviourist psychology' (Schapere 1981: 104).[2]

Banner criticizes his own discipline as a philosopher of religion: 'What ought to surprise us about lingering positivist prejudices and inclinations in the philosophy of religion and within theology is that logical positivism was fairly early on seen to be, as Ayer put it, pretty well false.'[3]

The next 'ism' we turn to is Sir Karl Popper's falsificationism. Popper was not a member of the Vienna Circle of logical positivists, but had close links with it. He forthrightly rejected the principle of verifiability in his book *The Logic of Scientific Discovery* (1934; Eng. trans. 1959), and condemned the attempt to put forward any general criterion of meaning. He aimed rather for a criterion to distinguish statements that are scientific from those that are not. As far as he was concerned, a statement may fail to pass his particular test for being scientific without implying that it was meaningless, even if it was metaphysical. His key criterion for establishing a test for scientific statements was falsifiability, leading him to insist that hypotheses formulated by scientists should be clear and precise enough to indicate how they could be falsified. For Popper, a scientific hypothesis must be clearly capable of falsification in order to be acceptable as a good hypothesis. In other words, a hypothesis should be formulated in such a way that it could be dealt a mortal blow by appropriate experiments and observations. This turned on its head some commonly accepted canons of scientific thinking: scientists are more concerned to justify their theories than to look for ways of falsifying them with counter-evidence. In popular belief, they also tend to concentrate on establishing new truths by observation and experiment; their aim is to make these as securely based as possible.

Popper believed that his account of how science progresses threw light on the unsolved problem of induction and how it works. The problem here is that, while we generally believe that additional favourable evidence for our hypotheses increases their credibility, as far as philosophers were concerned no satisfactory answer had been found to Hume's argument that reasoning from the particular to the general, or from past to future experience, cannot be justified. As Ayer put it, 'On this point Popper does not dissent from Hume, but he tries to cut the knot of the problem by denying that induction features in scientific method. There may be cases in which we are led to entertain a hypothesis by generalizing from experience, but how we come by our hypotheses is of no importance, except perhaps to psychologists. What matters is whether they are testable and how they meet the tests' (Ayer 1946: 133).[4]

Ayer's criticisms of Popper are somewhat unfashionable. For example, Lakatos has written: 'The great attraction of Popperian methodology lies in its clarity and force. Popper's deductive model of scientific criticism contains empirically falsifiable spatio-temporally universal propositions, initial conditions and their consequences. The weapon of criticism is the *modus tollens*: neither inductive logic not intuitive simplicity complicates the picture' (Lakatos 1970). Notwithstanding, Lakatos proposed his own variant, for what he called 'scientific research programmes': 'According to my methodology the great scientific achievements are research programmes which can be evaluated in terms of progressive and degenerating problem shifts; and scientific revolutions consist of one research programme superseding (overtaking in progress) another . . . this methodology offers a new rational reconstruction of science' (1970: 115).[5]

In rejecting verificationism, Popper argued that the view that scientific theories are falsifiable gives science its intrinsic virtue and strength. Popper contrasted this interpretation with what he took to be the older assumption that scientific laws are justified by induction. The crux of the problem arises when one realizes that there is no entailment of future behaviour from past evidence. For example, the sun rises every morning and we all have direct experience and knowledge of it, but this does not imply (logically) that the sun will rise tomorrow. While induction is certainly a habit that we depend upon day by day, induction may have no more philosophical status than a useful habit. The problem is that this account of science, and of knowledge in general, lends weight to the popular belief that religion and science are very different things; a religious worldview, while it may not be dismissed as meaningless, nevertheless does not have the quality of a theory which entails precise predictions which we can test in the way that Popper required. For example, we could predict that one day God will demonstrate that he has forgiven the sins of those who have repented and trusted in Christ and that hence there will be a last judgment, but such predictions are so remote as to make them unfalsifiable in the normal sense. Again, predictions from some of the Old Testament prophecies could be used to demonstrate our belief that they are indeed fulfilled in the life and death of Christ. But they would not really be deductions from theories in the way that Popper was arguing.

Popper was concerned with the testing of hypotheses which are ultimately always tentative. But religious beliefs are not essentially tentative hypotheses. Which martyr would have suffered and died for a tentative hypothesis? There is an element of trust and commitment in religious belief which is not part of tentative hypothesis testing. Instead of jumping on the bandwagon of Popper, the theologian would be better advised to join philosophers of science who have pointed out that many, and maybe all, theories ultimately fail Popper's trial by prediction.

A major weakness of Popperian reasoning lies in his treatment of sciences based on past events, such as evolution or astronomy. For example, Darwin's theory of natural selection may properly be regarded as a brilliant attempt to synthesize a mass of previously unordered evidence. Such a theory does not produce precise predictions in any straightforward sense. Indeed, for a theory to yield precise predictions we must know the initial conditions to which the theory is being applied, because theories do not predict in a vacuum. But Darwin was concerned with coherence, not prediction. In this sense at least, theism is not entirely dissimilar to Darwinian evolution!

Popper's particular philosophy of science gives an unsatisfactory account of the nature of science as a whole, though it does highlight one important aspect of scientific method. This has been voiced clearly by Harvard philosopher Hilary Putnam: 'The heart of Popper's scheme is the theory-prediction link. It is because theories imply basic sentences in the sense of "imply" associated with deductive logic – because basic sentences are *deducible* from theories – that, according to Popper, theories and general laws can be falsifiable by basic sentences. And this same link is the heart of the "inductivist scheme". Both schemes say: "Look at the predictions that a theory implies; see if those predictions are true" ' (Putnam 1981: 64). This is the crux of Putnam's unease:

> In a great many important cases, scientific theories do not imply predictions at all . . . [For example] The law of universal gravitation is not strongly falsifiable at all; yet it is surely a paradigm of a scientific theory. Scientists for over 200 years did not falsify universal gravitation; they derived predictions from universal gravitation in order to explain various astronomical facts. If a fact proved recalcitrant to this sort of explanation it was put aside as an anomaly (as in the case of Mercury). Popper's doctrine gives a correct account neither of the nature of the scientific theory nor of the practice of the scientific community in this case . . . All this is not to deny that scientists do sometimes derive predictions from theories . . . if Newton had not been able to derive Kepler's laws, for example, he would not have even put forward universal gravitation. And even if the predictions Newton had obtained from universal gravitation had been wildly wrong, universal gravitation

might still have been true: the auxiliary statements in conjunction with the theory might in his case have been wrong. Thus, even if a theory is 'knocked out' by an experimental test, the theory may still be right, and the theory may come back in at a later stage when it is discovered that the auxiliary statements were not useful approximations to the true situation. Falsification in science is not more conclusive than verification (Putnam 1981: 68).

The third of our 'isms' is that associated with Thomas Kuhn. His starting-point was that much of the philosophy of science simply did not fit the history of science. According to both positivism and falsificationism, the relentless application of their methods would guarantee the steady growth of science, with disagreement easily and quickly settled; but that is not the way that things really happened. There were frequently long periods of conflict in which groups of scientists disagreed with each other (for example, Copernicus and the Ptolemists among the astronomers, uniformitarians and catastrophists among the geologists, neutralists and selectionists among the evolutionists, and behaviourists and cognitive psychologists among the psychologists). The virtue of Kuhn's account is that it helps us to understand such extended disputes. In fact, when it was first expounded it was so good at explaining disagreement that it was difficult to discover how agreement ever occurred! For Kuhn, scientific discovery was not highly disciplined, rule-governed and cumulative, but was characterized by the occurrence of what he described as revolutionary 'paradigm shifts'. Paradigms were the generally accepted research methods and traditions at any particular given time; taken together, these combined the methods, generalizations and models shared by particular groups of scientists.

A good account of Kuhn's views is provided by Dudley Shapere (1981). Although he is primarily concerned with comparing Kuhn's views with those of Feyerabend, he also describes Kuhn's approach and its shortcomings. His argument is that Kuhn's criterion for accepting one paradigm as better than another was based on the notion of progress within the same paradigm, that is, within normal science. There, 'progress' consists of further articulation and specification of the paradigm within that science. The trouble comes when we ask how we can say that 'progress' is made when one paradigm is replaced by another. According to Kuhn, 'The differences between successive paradigms are both necessary and irreconcilable' (Kuhn 1970). A paradigm change entails 'changes in the standards governing permissible problems, concepts, and explanations' (Kuhn 1970), so what is metaphysics for one paradigm tradition may be science for another, or *vice versa*. It follows that decisions to adopt a new paradigm cannot be based on good reasons of any kind, factual or otherwise; quite the contrary, since what counts as a good reason is determined by the decision. Despite Kuhn's qualifications of this as extreme relativism, the logic of his position is clearly that the replacement of one paradigm by another is mere change, and not accumulation. Since they are 'incommensurable', two paradigms cannot be judged according to their ability to solve the same problem, deal with the same facts, or meet the same standards. 'The problems, facts, and standards are all defined by the paradigm, and are different – radically, incommensurably different – for different paradigms' (Kuhn 1970).

Putnam comments:

> Kuhn's most controversial assertions have to do with a process whereby a new paradigm supplants an old paradigm. Here he tends to be radically subjectivist (overly so, in my opinion): data, in the usual sense, cannot establish the superiority of one paradigm over another because data themselves are perceived through the spectacles of one paradigm or another. Changing from one paradigm to another requires a 'Gestalt switch'. The history and methodology of science get rewritten when there are major paradigm changes: so there are not 'neutral' historical and methodological canons to which to appeal. Kuhn also holds views on meaning and truth which are relativistic and, in my view, incorrect (Putnam 1981: 70).

Critics of Kuhn's view have pointed out that his account of science is essentially a sociological account of the nature of knowledge. Lakatos has gone so far as to describe it as an example of mob psychology among groups of scientists. It leaves us with an unsatisfactory account of the nature of science derived from a particular interpretation of the history of science. We are informed about why scientists disagreed, but given little evidence to guide us to understand why they agree from time to time. The fact is, however, that scientists do reach agreement, and may make major theoretical reconstructions of a whole field, leading a whole community of scientists to gather round and accept a new paradigm radically different from the previous one. The shift from Newtonian to Einsteinian physics would be one such shift, and the adoption of a broadly Darwinian approach to biology would be another. In these cases there was no significant and prolonged dissent, which ought to have been necessary and widespread on Kuhn's interpretation. Kuhn sought to answer his critics in *The Essential Tension* (1977), arguing that he never meant to deny a logic of justification, but only to deny that such a logic necessarily resembles a mathematical means of justification.

Another point where Kuhn's views fail is that if scientific progress is the choice of a mob based on mob psychology, or any other sort of sociological explanation, it is remarkable that science has been so outstandingly successful in applications on every hand, whether in medicine, engineering, geology, biochemistry, pharmacology or any other field. Such successes could hardly be predicated upon a purely social construction of reality. The fact is that success comes from scientists' discoveries about how the world is, rather than from sociological forces. From this it follows that Kuhn's account of science, if it is intrinsically flawed, cannot have the implications for theology which some have suggested. Michael Banner, for example, has argued that a Kuhnian account seems to be implicit in some of the writings of Don Cupitt: 'Generally speaking, this essentially sociological account of the nature of knowledge appealed to that breed of theologian which is hard to distinguish from the atheist or agnostic' (Banner 1990: 23).

Notwithstanding, Kuhn's views retain a wider influence. For example, Andrew Pickering has argued that scientific knowledge is a cultural construct, and 'there is no obligation upon anyone framing a view of the world to take account of what 20th Century science has to say' (1984). This has been highlighted by the widely publicized hoax by Alan Sokal, in which he assembled a virtually random set of clichés in an article which was accepted for publication in the journal *Social Text.* Commenting on this, Kurt Gottfried and Kenneth Wilson believe that while Sokal's victims may be a naïve fringe group, there remains a widespread, more sophisticated and troubling view of scientific knowledge championed by those whom they call the 'Edinburgh' school of sociology, a group which consciously look to Kuhn as an inspiration. They suggest that 'The "strong programme" Edinburgh sociologists contend that the knowledge produced by the natural sciences is a cultural construct essentially equivalent in its attributes to the knowledge produced by less tractable pursuits in which reproducible phenomena under carefully controlled conditions either do not exist or are of little interest. In making this case, its proponents dismiss or ignore a large body of concrete evidence that contradicts this contention. No view of the world is sound if it ignores the steadily improving predictions of twentieth-century science' (Gottfried & Wilson 1997: 545).[6]

We conclude that only a realist philosophy of science will be productive. In other words, the scientific enterprise presupposes that its efforts approximate to an accurate account of truth, of the way the world is. It is not that there is one clear-cut method of operation, but rather that our reasons for preferring one theory over another and for saying that one theory is more true than another turn out to be of a rather informal character. A theory is accepted because it is the best explanation of data or because it makes better sense of evidence than other theories, not because of any formal logical analyses, be they positivist or falsificationist. This in fact is the way that scientists actually argue for their theories in most instances. Frequently they have reasons which are more convincing in one way or another than logical proofs. This in turn means that attempts to make a hard-and-fast demarcation between science and religion, along the lines of positivists or falsificationists, are significantly weakened. There is, therefore, no reason in principle why it should not be possible to mount substantive arguments for theism. The sort of science that we want is that described by Paul Gross and Norman Levitt: 'Reality is the overseer at one's shoulder, ready to rap one's knuckles or to spring the trap into which one has been led by over-confidence, or by a too complacent reliance on mere surmise. Science succeeds precisely because it has accepted a bargain in which even the boldest imaginations stand hostage to reality. Reality is the unrelenting angel with whom scientists have agreed to wrestle' (Gross & Levitt 1994: 234). John Polkinghorne adopts a similar position, which can be called 'critical realist'.

Worldviews and pictures

We have already seen something of the early influences on the development of the scientific enterprise. We noted that large parts of the ancient scientific heritage were rejected in the sixteenth and seventeenth centuries, but science continued to benefit enormously from its logic, mathematics and occasional emphasis on experimentation. Christians were involved in a significant way in the early development of modern science. As an example of this effect, Hooykaas has contrasted the elaborate scientific world picture of the Greeks, which was a foundation for certain scientific disciplines such as astronomy and optics, with the way in which the Bible built up no single, comprehensive world picture and contains no scientific data which could sensibly serve as a basis for further scientific development. He believed this was the reason that the Hebrews made few contributions of lasting value to the material content of the sciences. What was more important was the difference in the general attitudes to nature held by the Greeks and those latent in the Bible, and the ways in which these attitudes conferred a lasting legacy on the two traditions. For the Greeks, this involved a rational investigation of nature, using logic, mathematics and some observation; but what was the basic attitude or worldview in Scripture? Hooykaas suggested that the correctives later found for the weaknesses of medieval science (a 'dedeification' of nature, a more modest estimation of human reason, and a higher respect for manual labour) were latent, although neglected, in the biblical tradition. 'If so,' he argued, 'we ought to be able to identify some general trends of thought in the Bible which could exert a healthy influence on the development of science, and it might be that in the 16th and 17th centuries these managed to overcome the shortcomings of the Greek attitude' (Hooykaas 1971).

Van Till (1989: 11–18) defines a worldview as a set of fundamental beliefs concerning the ultimate nature of reality:

Ordinarily it is expressible as a set of propositions concerning the identity of and inter-relationships among God, mankind and the rest of the world. A worldview is normally all-inclusive, concerned with both the physical and the non-physical, both the immanent and the transcendent. In so far as it is concerned with the physical universe, a worldview incorporates answers to questions about the physical world's status in relation to deity, about the ultimate source of the world's existence, about the identity of the agent that governs the world's physical behaviour, about the value of the world and its inhabitants and its component parts, and about the purpose of the world's existence and goal of its historical development.

Thus a worldview is primarily concerned with the ultimate nature of reality, and is a set of beliefs that produces a framework of meaning for interpreting life as a whole. For a worldview to be labelled 'Christian', it must have as its central focus God whom we know in Christ: the triune God – Father, Son and Holy Spirit – as revealed in the Scriptures as our creator, redeemer and comforter.

A second characteristic of a Christian worldview for Van Till derives from a specifically theistic doctrine of creation, involving an assertion that there is only one God who is the creator of everything. As such, the creation (that is, the entire Universe and all its inhabitants) is utterly dependent on God for its moment-by-moment existence, for its governance, and for the value it possesses and the purposes for which it was made. This, by implication, underlines the essential distinction between the creator and creation. Such a worldview contrasts radically with most other worldviews, including that of Israel's neighbours in Egypt and Canaan, who saw a world filled with many gods, frequently capricious and undependable. Their pantheons often included deities who appeared as parts of the creation: the Sun, the Moon, the stars and other 'powers' which impacted upon ordinary human life.

A third component of a Christian worldview is the way in which it differs from the metaphysical or philosophical naturalism often found in modern western societies. This secular view starts from the premise that the physical Universe is all there is; that there is no transcendent Deity. It implies that the Universe is self-existent and self-governing, and not dependent on any other being or power for its continuing moment-by-moment existence or its overall governance. In this sense, value and purpose, if there are any, have to be found within the Universe itself, so the Universe is seen as endowed with powers which are ordinarily ascribed to deity.

At this point it is important to distinguish between metaphysical and methodological naturalism. The latter is a scientific strategy which can commend strong Christian support. We do not continually have to reassert God's existence, because belief in his existence is already part of the historic foundations of the modern natural sciences. Methodological naturalism should be correctly understood as a useful approach to scientific work, without prejudicing its usefulness in other areas of life. Conversely, we need to recognize that natural science is constrained by methodological naturalism and does not pretend to produce answers to the ultimate questions. The question 'Why?' is not to be answered from within science, except to say that we have gone as far as our methods of natural science have taken us at this time. Christians should be comfortable with methodological naturalism in the natural sciences simply because God himself is sovereign over all life. God's works are manifest and open to study even by those who do not know him. God's incarnation affirms his capacity to be at home in the natural order. We should not force theological talk into science, because the natural sciences are necessarily incomplete (De Vries 1986).

Hooykaas distinguishes world pictures from worldviews. He defines a 'world picture' as a set of particular concepts about the contents and behaviour of the physical world. In this sense, a world picture, unlike a worldview, is not concerned with ultimate matters of religious import. A world picture limits itself in its scope to questions about the properties, behaviour and formative history of the physical Universe and its constituent parts. As Van Till has put it: 'To ask if the earth revolves around the sun is a question regarding a world picture; but to ask if the sun is a divine being is a worldview question' (1989).

Van Till describes a number of features of world pictures. He points out that our intellectual and cultural history includes a diversity of world pictures. The medieval world picture was essentially hierarchical and largely immutable – ideas derived from Plato, Aristotle and Ptolemy. In it, Earth, where humankind lived, was central in the Universe and was immovable; it was surrounded by a hierarchy of planetary and heavenly spheres. Likewise, creatures were ranked from the lowest forms of terrestrial life to humankind, and on up to angels, who were just below God himself. Such an ordering was deemed to be perfect and therefore unchangeable. Since the medieval world picture was so static, the idea that creatures might move from one rung to another on the hierarchical ladder was anathema to the medieval mind.

But world pictures are not in themselves unchangeable. Copernicus, Galileo and Newton all brought about revolutionary changes in prevailing world pictures: Copernicus and Galileo replaced Ptolemy's geocentric cosmological picture with a Sun-centred solar system; Newton demonstrated that the celestial and terrestrial realms behaved like each other. Today our world pictures rest primarily on the results of modern experimental and observational data (a method which has turned out to be remarkably fruitful in investigating the properties, behaviour, and so the formative history of the physical Universe). Our scientifically informed world picture presents us with a coherent and continuous functioning Universe, both in its day-to-day operating and in its historical development. This connects with our worldviews.

> Expressed in the language of the Christian worldview, we might say that God's creation is apparently characterized by an internal economy that has no gaps or deficiencies that need to be filled intrusively from outside of the creaturely realm. God has chosen, it seems, to constitute and govern the creative world, in such a way that the creator need not come down and act in a creaturely manner to fill in for gaps and deficiencies in his own handiwork . . . This is *not* to say, however, that the redeemer God does not act in extraordinary ways as he confronts us personally and redemptively in human history (Van Till 1989).

The question then remains how and in what way our present efforts to formulate a Christian worldview conform with reality as we find it in the created Universe. Van Till (1989) suggests three key features. First, it will lead us to reject any world picture that is not consistent with the crucial distinction between the creator and the creature. Hooykaas made the same point: 'There is a radical contrast between the deification of nature in pagan religion and, in rationalized form, in Greek philosophy, and the dedeification of nature in the Bible . . . Thus, in total contradiction to pagan religion, nature is not a deity to be feared and worshipped, but a work of God to be admired, studied and managed . . . The denial that God coincides with nature implies the denial that nature is god-like' (1971: 9). Secondly, to have a Christian worldview will lead us to reject any purely rationalistic approach to a world picture. Rational argument based solely on so-called self-existent principles, ideas or necessities is incapable of deducing the character or structure of the world in which we live. Thirdly, we shall at all times respect and acknowledge God's freedom to choose how the creation is made up, and therefore how it behaves and is ordered today, as well as specifying its formative history. God creates freely, and what we may discover about the structure and development of creation is simply the beginning of understanding his unhindered choices.

In principle, many world pictures could be compatible with the Christian worldview. We believe that God expects us to use the talents that he has given us to work towards discovering the particular character of the physical Universe; to this end the empirical approach has shown itself to be well suited to the task. It is this that the Christian scientists at the beginning of the sixteenth century saw as their high calling. Our commitment and calling are to open our minds to the results of empirical study and the results that come from careful scientific investigation. The more we learn, the more we may give thanks to the creator of the heavens and the Earth and of every other living creature that we find on the face of the earth.

Paths to knowledge: scientific and others

David Myers writes at the beginning of the most widely used textbook of psychology in North America: 'Psychology is a science that seeks to answer all sorts of questions about us all: how we think, feel, and act.' But he goes on to warn his student readers:

> If you ignore psychology's limits and expect it to answer the ultimate questions posed by the Russian novelist Leo Tolstoy in 1904 – 'Why should I live? Why should I do anything? Is there in life any purpose which the inevitable death that awaits me does not undo and destroy?' – you are destined for disappointment. If instead, you expect that psychology will help you better understand why people, yourself included, feel, think, and act as they do – or if you enjoy exploring such questions – then you should find the study of psychology both fascinating and applicable to life (Myers 1986).

It would be perfectly proper to substitute 'science' for 'psychology' in this passage, and thus make a general statement about the scope of the scientific enterprise. Myers is reminding us that if we believe that natural science is able to answer every meaningful question about reality, then we are making an unjustified and nonsensical assertion. Such an assumption is a philosophical or religious claim, based on a metaphysical belief usually described as scientism; it does not come from science itself. To assert that science is the only pathway to sound knowledge about reality is to espouse scientism. There are certainly scientists who do make such claims, but they are speaking of their personal beliefs and not as a voice of the scientific enterprise.

Apart from those who embrace scientism (the assumption that natural science provides the only proper elements for philosophical enquiry), most of those engaged in the scientific enterprise assert that the questions they seek to answer are concerned with the properties or behaviour of the objects that they study, together with an interest in tracing the formative history of the objects concerned. For example, neuroscientists want to know the physical and biological properties of the nerve cells they study. What is the structure of the basic unit of the nervous system? How do nerve cells behave individually and in groups? How do they transmit information, and how do the processes of excitation and inhibition work? How has the brain come to be the way it is? Can studying the brains of non-human primates and other vertebrates help? In other words, can evolutionary processes tell us something of the formative history that gave rise to the human brain? The same sorts of questions arise in other disciplines, and it is easy to think of the questions that have to be asked.

When we take together the answers from such questions, we believe that we gain an increase in our knowledge of the inherent intelligibility of the Universe in which we live and of which we ourselves are a part. It is important to add the qualifier 'inherent', because 'the object of investigation by the natural sciences is not all there is. There is more to reality than the physical alone' (Van Till 1989).[7] But this means that there are other areas of knowledge with which the scientist does not deal: abstract ideas, concerns with transcendent beings or concepts such as beauty and truth, moral principles for right and wrong, and spiritual beings and their possible actions. Discussion of transcendent beings or transcendent relationships lie outside science. The scientific enterprise is not designed or competent to discuss the relationship of the world to any divine creator. Questions about the relationship of the Universe to God must be directed elsewhere. In general, the self-imposed silence of the scientific enterprise in such matters is honoured by practising scientists, whether they are non-theists or theists. Both seek to resist the temptation to use science to give warrant for (in the sense of proving) their own particular religious perspective as against another.

The distinction between the domains of inherent intelligibility and transcendent relationships is central to a proper understanding of the relationship between the scientific enterprise and Christian faith. Astronomers talk about the origin and evolution of planets, stars and galaxies, but their concern is with the processes whereby these celestial objects arose from earlier structures. Cosmologists speaking about the origin of the Universe focus on understanding the processes by which the present state of affairs has come about. Any question of ultimate origin is outside their competence as scientists. To ask, 'What is the ultimate origin of the Universe?' is to ask about the very existence of everything. To ask that sort of question is not only to ask about the beginning of existence, but even about the existence of time itself. Questions concerning the origin of the Universe are real and proper ones, but the important point is that answers to them take us outside natural science and into the domain of metaphysics or religion.

A second distinction is between the behaviour of the physical Universe and fundamental questions concerning its governance. When scientists investigate the behaviour of the natural world, they are necessarily concerned with empirically accessible processes. Thus physicists try to understand the behaviour of physical systems and their interactions in terms of fundamental forces related to the physical properties of matter. Biologists try to understand the behaviour of living systems in terms of the structure and behaviour of cells as well as in terms of interaction between the organism and its environment. But we are all interested in looking for more comprehensive sets of interrelated patterns; these form our scientific theories or theoretical models, or possibly laws of nature. Although people at times talk about 'the laws of nature that govern the behaviour of physical systems', such an expression is really quite vacuous. The laws of nature are our descriptions of the patterns that we have detected, and descriptions have no power whatever to govern; understanding of governance lies outside natural science. Within the Hebrew–Christian tradition, God is the governor, the upholder and sustainer of all things. What we readily call the laws of nature are our description of the patterns of the way this divine governance manifests itself. Questions of ultimate origin and governance are highly important questions, but they must be directed towards what each one of us sees as the ultimate source of answers to our religious questions, and not at the scientific enterprise.

Values and the scientific enterprise

As we have seen, to suppose that there is a set of self-evident or rigidly applicable rules for establishing the truth or falsehood of a particular theory once and for all is an inadequate and inaccurate account of the scientific enterprise. In the past few decades, historians, sociologists and philosophers of science have converged on a more realistic assessment of the way science actually works. This means that the scientific enterprise is dependent on values and judgments of individual scientists. If we want to understand how scientific theories are developed and evaluated, we have to study the scientific enterprise as a human enterprise. It is essentially the activity of a community of individuals who, having acquired certain knowledge and training in particular skills, go on to make judgments about the adequacy of particular scientific theories in the light of empirical data. There is no single authoritative source which spells out for the whole scientific community what are or are not acceptable criteria for scientific investigation and theory evaluation; there are, rather, unwritten rules latent in the scientific community and its practices. Van Till (1989) has summarized the values operating in the scientific community as 'matters of competence', 'integrity' and 'sound judgment'. By 'matters of competence' he means that a good scientist must dedicate himself to acquiring the basic knowledge and skills that are required to carry out empirical, analytical and theoretical operations effectively. A long training is required in many scientific disciplines, and those who fail to acquire and obey such training are understandably rejected as incompetent by their scientific colleagues. As far as 'integrity' is concerned, we have to depend upon the professional integrity of other members of the community of which we are a part. We may debate the significance of a scientific report and we may raise questions about the power of its arguments, but we take its integrity as given. Sadly, this has been shown in recent years to be an assumption that we cannot always make. 'Sound judgment' involves applying the criteria normally used to evaluate the scientific merit of a particular theory, including predictive accuracy, internal coherence, external consistency, unifying power, fertility, and simplicity. There are, of course, other criteria; some mathematicians and theoretical physicists set great store on the beauty and simplicity of a theory. The problem is that such value judgments tend to depend more on preconceptions from philosophical commitment or religious beliefs than the disciplines intrinsic to science itself.

Such an approach stands the scientific enterprise on its head and must be resolutely avoided. The goal of natural science is to gain knowledge, not to reinforce preconceptions. The purpose of empirical research is to discover what the physical world, as an intelligible system, is really like, not to verify its conformity to our prejudices. And the aim of scientific theorizing is to describe the actual character of the universe, not to force its compliance with our preconceived requirements . . . Science held hostage to any ideology or belief system, whether naturalistic or theistic, can no longer function effectively to gain knowledge of the physical universe (Van Till *et al* 1990: 149–150).

Conclusions

Historically, Christians have been intimately involved in the scientific enterprise, partially at least as fulfilling a call to care for the created order as stewards.

The contributions of philosophers of science have been considerable, although it is important to remember that science was practised effectively before they undertook detailed analyses of the nature of the scientific method. These analyses of the logic of scientific discovery have given us new insights into what scientists have intuitively been doing in many cases. Whether in the mould of the positivists, the falsificationists or the Kuhnians, their approaches have brought out important aspects of the scientific enterprise and the logic of scientific discovery. But they have become self-defeating when they have claimed that only this or that analysis of the logic of scientific discovery *really* represents *the* scientific method. A. J. Ayer commented on falsificationism that 'Popper gives a luminous account of at least one form of scientific procedure, but the basis of his system is insecure' (1946: 134).

The Christian worldview positively encourages involvement in the scientific enterprise; but we must beware of some world pictures, necessarily temporary, because they can readily lead to the abuse of Scripture and become a barrier to the development of science.

Chapter 4

Explanations, models, images and reality in science and religion

What do we mean, and what is being claimed, when we say that we are able to give a 'scientific explanation' of some feature of the natural world? Since we use models and pictures both in our scientific thinking and in our religious discourse, how do the usages of these terms compare and contrast in the different domains? Within science there are various modes and levels of explanation which are not necessarily regarded as logical alternatives or competitors. In this chapter we argue that we misunderstand the nature of a scientific explanation if we regard it as a logical alternative to, or a necessary competitor with, the various biblical portraits that we find concerning, for example, creation. We raise the question of whether some of the enthusiasm for demythologizing in theology comes from a mistaken understanding of models in science.

The nature of explanation in science

The central question we shall concern ourselves with in this section is 'What do we mean when we say that we give a scientific explanation of an event or a phenomenon?' We have already touched upon this, but it is an important key to the whole understanding of the relationship of God to his world, so we must examine it in detail. Its importance lies in the fact that there has, from time to time, been considerable confusion about what is meant by 'scientific explanation', not only in the minds of some Christians, but also of some scientists. The former have, for example, at times regarded scientific explanations as being in direct competition with, and therefore as directly threatening, statements which have traditionally been made in religious language about the activity of God. Richard Dawkins is a recent proponent of the latter. We have to make clear what is and what is not being claimed when we say that we can offer a scientific explanation of some phenomenon in the natural world, before discussing questions about how to relate scientific and religious statements.

Some widely held views

We begin by briefly surveying a range of different meanings which have been given by scientists themselves and by philosophers of science to the idea of explanation in science. Some have argued that to explain is logically equivalent to predicting, and nothing more. Those who take this view argue that the scientist's main objective is to show that an event studied can be subsumed as an instance of a general law, which can then be stated. This approach means that any event, whether in the past or in the future, should be deducible from the application of the law to all the relevant information about the conditions existing before the event took place. Others have challenged this view on the grounds that many theories which are generally accepted as scientific explanations, such as the theory of natural selection, would be held by very few to have been capable of predicting, in this case, the course of evolution. It should also be pointed out that the converse does not automatically apply, and that, while an explanation may lead to the prediction of future events, nevertheless prediction may not depend on explanation.

The fact that we can often predict future events with no real understanding of *why* we can predict them explains much of our motivation for scientific research. If explanation is logically equivalent to prediction, one could presumably further argue that scientific explanation becomes unnecessary, provided that we can make the necessary prediction without our scientific knowledge. Thus the search for an explanation includes a concern for identifying relationships of cause and effect, while prediction does not necessarily do so. It would be very difficult to find any scientist who would be content merely to predict, without also being able to understand the mechanisms operating.

Usually, explanations become satisfying to a scientist only when they can be shown to be derivable from theories. This not only gives some explanation or understanding of the events being studied, but also suggests their applicability to new types of phenomena not yet understood. Predictive laws alone must always be dissatisfying to the scientist, until such time as we gain insight into the theoretical structures which can account for their consistent success. No doubt it is for this reason that writers such as Toulmin (1961; 1965) and Hanson (1958) have insisted that one of the central aims of the scientific enterprise is not simply to be able to explain isolated phenomena, but rather to gain an increased understanding or intelligibility of the whole pattern of events that are being observed. Toulmin, for example, points out that the ancient Babylonians could make astronomical predictions, but had very little understanding of why their predictions turned out to be correct more often than not. Others have argued that the difference between explanations and descriptions is not always clear, and that it may sometimes be the case that an explanation is nothing more than a comprehensive description of an event. The point of a good explanation, however, is that it tends to make use of ideas at a different logical level than the event being described and that it therefore includes that which is being explained in a more general and comprehensive system.

The nature of scientific explanations really serves to focus on the much more general and important question of how we view the relation of scientific statements to the reality with which they purport to be concerned. We have shown that the empiricist tradition of Bacon, Hume, Mill and modern practitioners includes those who place the most emphasis on the observational side of science. Positivists argue from this that scientific explanations, including the concepts and theories that go with them, are really nothing more than concise summaries of the data being studied. Such a view, however, is difficult to defend, since the point of a scientific theory is that it introduces new conceptual terms which are not the same as the observations of which they are designed to make sense. The important thing about a scientific theory, in so far as it constitutes an explanation of the phenomenon being studied, is precisely that it makes use of ideas at a different logical level, which have a greater comprehensiveness and generality than the phenomenon being observed.

Another approach stemming from the developments in logical positivism, mainly within British philosophy which led to linguistic analysis, is often referred to as 'instrumentalism'. This particular view has been common among philosophers of science. Its main point is that scientific explanations are 'maxims or directions for the investigator to find his way about', whereas theories are 'techniques for drawing inferences', useful primarily for making predictions (Toulmin 1953).

On this view, explanation is not to be judged by its truth or falsity, but by its usefulness in 'making accurate predictions', 'directing further experimentation' and 'achieving technical control over the environment' (Barbour 1966). This would mean that scientific laws and theories are invented by scientists, not discovered as properties of that which is being observed. In practice, however, most scientists do in fact regard the evidence which they gather in their experiments as counting for or against the *validity* of a particular theory, and they are not simply content to regard it as counting for or against the *usefulness* of that theory. It is not at all uncommon for two apparently contradictory theories or explanations to turn out to be useful in the sense of generating further work. The classical example of this is light, which can be explained both as particles and as waves. Nevertheless, we continue to seek to resolve the conflict between such explanations by exploring which of them corresponds most nearly with the real situation; in other words, which of them is the true explanation.

Some philosophers take a different view, and regard the explanatory theories proposed by scientists as largely structures imposed by the human mind upon the chaos of sense data which confront us and by which we are continually bombarded. Such a view leads to the almost complete neglect of the way in which the data guide the scientist in formulating explanations. It is thus at the other extreme from the positivist belief. While the positivists stress the controlling aspect of the data so as to neglect the participation of the knower, this view so stresses the participation of the knower that it ends up by neglecting the part played by the regularities in nature which first caught our attention.

Finally, there are those who restore the balance somewhat in that they assert that patterns in data, while they are to some extent imposed by us, nevertheless partly originate in existing objective relationships within nature itself. They would stress that the scientists' activities of discovery and exploration are just as important as those of invention and theory construction. Such an interpretation regards explanatory theories as in essence representations of what is the case in the world around us. This view was developed extensively by the philosopher and mathematician A. N. Whitehead (1926) in his so-called realist epistemology.

A Christian view of scientific explanation

Our own view is close to this kind of realist position. We believe that as Christians we are committed to maintain neither an extreme objectivist nor an extreme subjectivist position concerning the nature of our knowledge of the physical world. The Bible is explicit that such order and uniformity as we discover in the range and variety of the world around us are part of God's creation. Its being, including its capacities for patterned behaviour and its capabilities for achieving orderly structures, has been given by God and is sustained by God's continuing action as its creator. We believe that our Christian faith will engender within us the confidence that we are equipped with sensory and intellectual abilities which will enable us, by perceiving and studying the phenomena around us, to acquire the knowledge that we need in order to carry out the cultural charge given to us by God.[1]

From this point of view, physical knowledge in science, and the theories, laws and explanations which it summarizes, are all found 'to have essentially the character of a qualified relation extending within the created order' between the two poles, 'the intelligent subject, man, on the one hand, and the intelligible phenomena, the rest of nature, on the other' (G. J. Sizoo, quoted in Jeeves 1969: 48). This being so, any view which minimizes either of these extremes to the extent of failing to do it justice would be a distorted view.

From time to time in the course of our scientific activities, the emphasis on one or the other of these poles may legitimately be increased. Moreover, our experience tells us, and the history of science confirms, that in order to be good scientists we do not *need* to decide whether we accept one or other of the views outlined above, or even whether the pattern is in the events or only in our heads. What usually happens is that we notice regularities in nature, we collect our data, and then, in seeking to make sense of them, we develop our law-making, our theorizing and our explanations, all of which must continually be guided by the data. At other times our activity seems to be almost entirely that of sitting back and turning over in our minds the properties of the accumulated data and imposing one or another of a whole variety of possible patterns upon them, in an attempt to see which one best makes sense and best fits with our previous experience. One thing, however, is clear: our explanation of reality is not identical with that reality, otherwise it would imply that as our theories, our laws and our explanations change with the development of science, so the reality which we are studying is also changing, and this we certainly would not believe.

We shall have more to say about how our explanations are related to the physical world they purport to explain. Perhaps we should make one general point about all discussions (including our own) of what we mean by explanation, law or theory in science, so that we do not let reason sit in judgment. While as scientists we can appreciate the worthwhileness of the activities of the philosophers of science, and of those who are interested in the logic of the scientific method, we can nevertheless get along perfectly well with our scientific activities *without* continually stopping to analyse what we are doing. Some philosophers and some scientists, of course, take a much more extreme view and assert that

> Among the obstacles to scientific progress high place must certainly be assigned to the analysis of scientific procedure which logic has provided . . . it does not try to describe the methods by which the sciences have actually advanced, and to extract . . . the rules which might be used to regulate scientific progress, but has freely rearranged the actual procedure in accordance with its prejudices; for the order of discovery there has been substituted an order of proof . . . it is not too much to say that the more deference men of science have paid to logic,

the worse it has been for the scientific value of their reasoning . . .
Fortunately for the world, however, the great men of science have
usually been kept in salutary ignorance of the logical tradition (Schiller
1955).

Levels of explanation

Anyone familiar with (say) the scientific literature on human and animal
behaviour is well accustomed to finding a variety of different explanations of
the same behaviour, without seeing them necessarily as competitors with
each other. For example, if one were studying a simple piece of learning by
an animal in a maze, it would be possible to explain the animal's behaviour
in terms of psychological concepts such as the amount of reward or of
behaviour reinforcement, or the intervals between the reinforcements, or
how many times the animal has been reinforced, or other similar concepts. It
would be equally interesting to study the animal from the point of view of
changes in the electrical activity in different parts of the animal's brain as it
learns (the sort of thing which has in fact been done using telemetry and
implanted electrodes). Yet another scientist may well be interested in the
changes in the biochemical activity of the brain in the course of such
learning. No scientist would think of placing these different explanations
alongside each other as competitors and of using one to exclude the others.

It is not at all uncommon in biology to find different kinds of explanation
grouped in terms of *modes* of explanation rather than *levels* of explanation.
The palaeontologist G. G. Simpson (1950), for example, distinguished three
modes of explanation commonly used by biologists: first, explanations which
are answers to the question 'How?' in terms of the mechanism involved, often
labelled as reductionist explanations; second, answers to the question 'What
for?', where one is looking for an answer in terms of function, referred to as
compositional explanations; and third, answers to the question 'How did this
come about?', that is to say, answers in terms of the formational history of
the organism.

The second mode of explanation identified by Simpson includes an
analysis of the adaptive usefulness of structures and processes, both for the
entire organism and for the species to which it belongs. Of particular interest
here are the evidences which come up repeatedly of apparently purposeful
behaviour displayed by animals. Ernst Mayr (1982) speaks of these as
instances of programmed behaviour, on the basis of analogy with computer
programmes. Others have sought to do without this kind of explanation by
talking in terms of goal-directed behaviour instead of purposive behaviour.
The third type of explanation, labelled historical explanation, is usually
inappropriate for strictly physical phenomena, but it is both appropriate and
necessary in biology. In genetics, for example, it is impossible to understand

the present frequency of different genes without considering the genetical processes acting upon previous generations.

The point is that any one of these levels or modes of explanation may well be *exhaustive* at a particular level or in one particular mode, but this does not mean that it is the only possible or indeed the *exclusive* explanation, which has to be involved for a full account of the phenomenon which is being studied. This is an important point to which we must return in detail later, since it has implications when we seek to relate scientific explanations to other explanations given more traditionally in different terms, including religious ones.

There are important differences between sciences, and these frequently surface in inter-disciplinary debates. In particular, biological sciences have to accept and allow for variation at all levels, whereas (in the words of Cambridge zoologist Carl Pantin) 'physics and chemistry have been able to become exact and mature just because so much of the wealth of natural phenomena is excluded from their study'. Pantin suggested that physics and chemistry should be called 'restricted' sciences, in contrast to the 'unrestricted' sciences of biology and geology (1968: 24). Marjorie Grene adds to this. She argues that although all sciences share a basic search for pattern in the events that are being studied, nevertheless the restricted sciences must add something if they are to do full justice to the subject-matter of their enquiry.[2] Even in areas of biology such as genetics, where physical sciences are increasingly taking over, this point is still important.[3]

This is an interpretation which is denied by doctrinaire reductionists like Richard Dawkins, for whom an organism is defined merely as a gene's way of ensuring its own survival. Grene faces this criticism head on. She points out that, in the exact sciences, an explanation ideally provides some sort of qualitative or quantitative formulation, from which statements about a certain range of phenomena can be deduced, whereas

> The geneticist's recognition of a fruit fly stands in a different logical or epistemological relation to the theories of genetics from the relation, say, of the reading of the temperature or pressure of gas to the kinetic theory of gases. For over and above the recognition of pattern implicit in the grasp of data relevant to theories, biology demands the recognition of individuals, to which as its *raison d'être* it has continually to return (Grene 1966: 210).[4]

What are we to conclude from this discussion of an apparently rather specialized topic within biology? First, that none of the various modes of explanation which have been suggested should necessarily be regarded as logical alternatives to or competitors with each other. Nor, we insist, should any of them be regarded as necessary competitors with the various biblical affirmations which we find concerning creation. Certainly no scientist has a logical basis for insisting that scientific explanations provide grounds for denying the activity of God in sustaining his creation, or for disproving his existence. Indeed, the consideration of a variety of modes of explanation in biology, as outlined above, should perhaps help us to see the dangers of speaking over-dogmatically, as some non-theists do, about what a scientific explanation does and does not imply about other types of explanation.

In the minds of Christians, scientific explanations have too often been regarded as in direct competition with assertions about the activity of God. The misunderstanding is, moreover, by no means one-sided, because there are also some non-believing scientists who maintain that a scientific explanation of a phenomenon thereby excludes the appropriateness of any other sort of explanation of the same phenomenon. A problem here is that the Bible contains elements of description, of explanations and of prediction; it will be our task in the following sections to see how we should relate these to different types of explanations given by scientists in their various disciplines.

Relating explanations, models and reality in science and religion

So far we have explored a variety of ways in which the notion of explanation is used within science itself. We have seen that there may be a whole cluster of possible explanations of a given phenomenon, all equally scientific, yet some will be based on macroscopic concepts and others on microscopic concepts. In some instances a higher-level (macroscopic) explanation may ultimately be absorbed completely into, or replaced by, a lower-level (microscopic) explanation. But, particularly in biology, this need not be so. The question now, however, is this: granted that we may categorize a wide variety of scientific explanations together as one group or type of explanation, namely the scientific, how are we to relate this type of explanation to the other type of explanation, particularly traditional religious ones? Is there any sense, for example, in which scientific and religious explanations must necessarily compete for our allegiance and acceptance?

There are several possible ways of approaching this problem which are improper for a Christian. For example, a scientist who is also a Christian has consistently to resist the temptation to seek the easy way out by compartmentalizing life, experience and thinking; in other words insulating scientific knowledge from Christian beliefs. Any solution that relates scientific knowledge to Christian beliefs must do full justice to all of one's experience as a scientist and a Christian. This point perhaps is sometimes not understood by our non-Christian colleagues, who fail to realize that our Christian experience is just as much part of our experience as taking measurements in laboratories. We could ignore our Christian experience only by being untrue to ourselves and behaving as spiritual schizophrenics.

One way of relating our scientific knowledge and our Christian beliefs, which had considerable vogue some years ago and still lingers, has been dubbed the 'God-of-the-gaps' approach. The name will always be associated with the Oxford mathematician Charles Coulson (1955). It can be illustrated with a quotation from theologian W. A. Whitehouse:

> I am myself inclined to think that the mystery of God's Providence lies deeper than the eruption into nature of such interferences [he is thinking of the possible control of matter by mind] and I am attracted by the fact that scientific explanations and predictions rest now on 'the law of great numbers'; that fundamental physical laws are statistical and not exact in the popular sense. Why this should be so is an interesting matter for speculation. It may provide a sufficient room for manoeuvre beneath the observable, regular processes, for the personal care of God to be actively exercised (Whitehouse 1952: 121).

The key phrase is 'room for manoeuvre', which implies that nature has things more or less tied up, but that there may be a few gaps in which God can still have his own way. And it is not only the theologians who at times have written in this way; so also have some distinguished scientists, such as a leading astrophysicist who wrote approvingly of 'the notion of God continually intervening, with deft touches now here, now there, to direct the material particles in the universe so as to conform to rationally deduced laws' (E. A. Milne 1952: 156).

Once again, we have the idea of God intervening from time to time in the gaps in the otherwise orderly running of the Universe to bring about his divine purposes. According to this view, one explains things scientifically as far as one can, and then one brings in God to explain what is still inexplicable at the scientific level. In this way one divides up the territory of investigation into those bits of nature that science can explain and that God cannot touch, and those bits where science has so far failed and perhaps God could be at work. The net result is that God is left with a steadily dwindling territory, shrinking with every new scientific discovery. This, of course, is not to suggest that science has as yet dealt with most of the ground to be covered. Anyone actively engaged in scientific research is all too aware of the tremendous area still to be explored. The point is that, however imperfectly the scientist understands any process studied, it would be advancing a non-Christian idea of God to suggest that God can be seen at work only in the bits of nature that continue to puzzle the scientist.

Divine upholding

We stressed in chapter 2 the importance of the biblical emphasis on the moment-by-moment divine upholding of the Universe. Some of the statements in the Bible have direct reference to the natural order: for example, our Lord's own words when he said that God 'causes his sun to rise on the evil and the good' (Mt. 5:45), or when he asserted that it is his Father and our Father who feeds the 'birds of the air' (Mt. 6:26). Neither of these claims states anything extra or contrary to a physical explanation of the movements of the planets or the way birds are fed. What is implied is that when we have finished analysing the movement of the Sun or the feeding behaviour of birds in physical terms, there remains a fresh sense to be made of the same pattern of events, if we are to do full justice to what is given to us.

This may be illustrated by MacKay's story of two people sitting on a cliff-top looking out to sea. One of them was a physicist, and a very enthusiastic one, who carried a good deal of his scientific equipment in the boot of his car. The two saw a light flash on and off out at sea. Our physicist boasted that, given a little time, he would be able to give a full account of the wavelength, the emission rate, the frequency and the various other characteristics of the light that was flashing. His friend became increasingly agitated, since in the distant past he had learned the Morse code and was dimly aware that the light flashes were also communicating a message. In fact, they were saying that the piece of cliff on which the couple were sitting was beginning to crumble and would shortly slide into the sea. One could reasonably expect the physicist to give a complete and exhaustive description in physical terms of all that was occurring at the light source, and yet this alone would leave out another, and in the circumstances extremely important, aspect of the same phenomenon, namely the purpose of the events which were being observed. The meaning and significance in this case were there for those who were able and willing to read it in a different way. The point is that before one reads two assertions about the same phenomenon as contradictory, one should be sure that they are not in fact logically complementary.

The history of science shows this well. In the 1800s the orthodox way of picturing light was in terms of waves spreading through space. The evidence for this was very convincing, and it seemed clear that the earlier view of picturing light as a stream of particles was wrong and should be abandoned. But the situation changed again when it was discovered that, in certain previously unexplored situations, light behaved quite definitely like a hail of tiny particles. Which was the true picture, the wave or the particle one? Only after a lot of hard thinking did it become clear that the correct answer was that both pictures could be valid; the two interpretations were not rivals, but complements. The lesson is that we cannot deduce contradictory conclusions by a proper use of the two experimental approaches to the nature of light because they represented answers to different kinds of question. This 'Principle of Complementarity', first enunciated for physics by Niels Bohr, is an analogy, and not in any sense a proof of the necessity of complementary Christian and scientific viewpoints. However satisfied we may be that the two pictures are compatible, only the facts of experience can convince us that both are necessary. We are dealing here with a logical point, not a scientific one, but it is one which is open to easy abuse and misunderstanding. For this reason we need to see clearly the conditions under which it can legitimately be used. Unless we do this, it could easily become an escape hatch which we use when we get into a tight corner in discussions concerning the relation of science and faith. MacKay drew attention to this danger:

> Whenever a new concept swims into philosophical ken there is a danger that it will be overworked by the Athenians on the one hand and abused by the Laodiceans on the other . . . Complementarity is no universal panacea, and it is a relationship that can be predicated of two descriptions only with careful safeguards against admitting nonsense. Indeed the difficult task is not to establish the possibility that two statements are logically complementary, but to find a rigorous way of detecting when they are not . . .
>
> A good deal of consecrated hard work is needed on the part of Christians to develop a more coherent and more biblical picture of the relationship between the two . . .
>
> But if once we recognize that at least most theological categories are not 'in the same plane' (in the same logical subspace) as most scientific categories, there is no longer any theological merit in hunting for gaps in the scientific pattern. Gaps there are in plenty. But . . . it would seem to be the Christian's duty to allow − indeed to help − these gaps to fill or widen as they will, in humble and cheerful obedience to the truth as God reveals it through our scientific discipline, believing that to have theological stakes in scientific answers to scientific questions is to err in company with those unbelievers who do the like (MacKay 1953: 163).

MacKay detailed the conditions under which two or more descriptions may legitimately be called logically complementary. These are (1) that they purport to have a common reference; (2) that each is in principle exhaustive (in the sense that none of the entities or events comprising the common reference need to be left unaccounted for); yet (3) they make different assertions because (4) the logical preconditions of definition and/or of the use (that is the context in which they are set) of concepts or relationships in each are mutually exclusive, so that the significant aspects referred to in one are necessarily omitted from the other.

MacKay also pointed out that nothing in the idea of logical complementarity excludes the possibility of a higher mode of representation which could synthesize two or more complementary descriptions; nor is it necessary that one description should be inferrable from the other. The notion of complementarity involves a warning not to relate such descriptions in the wrong way, by treating them as (a) referring to different things, (b) synonymous, (c) exhaustive, or (d) contradictory.

The somewhat negative point which arises from all this is that before religious and scientific statements could be debated as rivals, it would be necessary to establish that they are not in fact complements. It is also of course equally necessary to realize and to recognize that proof of complementarity would not establish that either account was true.

The uses of analogies, models and images in scientific and religious thinking compared and contrasted

We have already noted the dangers of assuming that particular analogies, models or theories that we use in science are the same as the reality they purport to help us to understand. We have also noted that developments in physics in particular have weaned us from the nineteenth-century idea that science was providing a *literal description* of an objective world. Such a naïve realism is certainly not tenable today. We now realize that our concepts are symbols which help us to deal with certain limited and restricted phenomena in order to achieve particular purposes. Concepts are not given to us in nature, but are devised and made use of in our human symbol systems. Not only are our mathematical symbols *not* the physical world, but also our mathematical laws do *not* cause the world to revolve. In short, physical laws do not cause physical processes to take place, and we must never use the term 'cause and effect' in science with this connotation. As Wittgenstein put it, 'At the basis of the whole modern view of the world lies the illusion that the so-called laws of nature are the explanations of natural phenomena' (cited in Jeeves 1969: 72).

We have seen how scientists begin by assuming that there is an objective reality which is available for study, that there are regularities in the behaviour of this reality which can be observed and recorded, and that, having carried out observations and recording, they may then be in a position to use analogies and models which may help them understand something of these regularities. An analogy in this sense is an observed or inferred similarity between two situations which have some properties, forms or functions in common, although others may be different. Such analogies help us in our scientific enquiry, in that they enable us to envisage patterns of relationships and to extend them to other areas and types of experience. An analogy becomes a model in our scientific theorizing when we use it to describe a phenomenon whose laws are already known and extend it in a systematic way to another one which is still under investigation. Our models may be mathematical models, so that there is a formal similarity in the equations representing the two phenomena; or they may be mechanical models, such as the billiard-ball model on which the kinetic theory of gases was based.

Frequently in the course of scientific discovery, 'analogies and models' have turned out to be extremely 'fruitful sources of scientific theories'. Dangers arise when failure to realize that models deal only with selected characters results in a tendency to over-extend them and to assume that all features of the analogy will be present in a new situation. When used in this way a model becomes a hindrance to further development. This has led some to take an extreme point of view and argue that models are merely 'temporary psychological aids in the formation of theories' (Barbour 1966: 159). The proper, pragmatic use of models has been described by philosophers of science such as Hesse, Toulmin and Nagel: if they are good models, they suggest further questions, which take us beyond the phenomena with which we began and hence enable us to formulate hypotheses which may also turn out to be experimentally fruitful.

Most writers on this topic seem to agree that the lesson which we should learn from the nineteenth century is that models must not be interpreted literally. An analogy is never totally identifiable with, nor is it a complete description of, a phenomenon; it is only a simplified comparison of limited aspects of the phenomenon with some other phenomenon. This has become increasingly clear as the value of mechanical and visualizable models has declined, particularly in quantum physics, where it is no longer possible to visualize the way the atom is represented by certain wave functions.

Good science must reject naïve realism, since that is a limited approach which fails to do justice to the part our intellect plays in the creation of theories, and it also fails to acknowledge the extremely selective nature of most theorizing. In our desire for an understanding of the physical world, we must be careful not to identify any particular model or theory which summarizes our understanding with what is being studied and understood. If we do, we can maintain the validity of our first assumption in the light of new evidence only by suggesting that reality is changing also. In short, we must always be on our guard against identifying the reality that we study with the analogies, models, images and theories which we use in summarizing our understanding of that reality.

Models, images and reality in religious thinking

As in science, so in religion we find ourselves using analogies, models, stories, thought-constructs and word pictures of various kinds in order to think about and to communicate ideas about spiritual reality. In science we make use of analogies in order to extend patterns of relationships which we have understood in one context to another area of experience which has similarities with it, and in the same way we frequently use analogies in religion to interpret our religious experience and to communicate ideas about God. The Bible is full of analogies of God, such as the king, the rock, the strong tower, light, or power.

Parables may be regarded as a special form of analogy in which a vivid story taken from everyday life portrays one central truth about some aspect of the relation between humankind and God. Even this can be distorted if we try to make every detail of the parable carry a meaning which is symbolic of some aspect of the God–humankind relationship. In general, parables are designed to communicate one main point, as for example when God is compared with a father or with a king. Much more could, of course, be said about parables if we were writing in a theological context.

We suggested earlier that when a set of analogies is systematically brought together in science, these analogies may be used to form a single, more comprehensive model. In biblical theology the central model for God is the human person. But we must be careful to ask which aspects of the model we should regard as important and relevant, and which are irrelevant and inapplicable. For example, the model of God as a person does not need to imply that God has hands and feet, even though from time to time we find passages which speak of God in this way, as for example, the statement that he was 'walking in the garden in the cool of the day' (Gn. 3:8). Moreover, we find particular kinds of personality portrayed to us as teaching different aspects of the personality of God: for example, God as the sovereign ruler, God as the just judge, God as the loving father, and so on. But all these are incomplete, because there is one supreme model of God, one point at which the model and the reality have become one, namely the person of Christ himself.

Just as there are dangers in the use of models in science, so there are dangers present in the use of models in religion. If we expect so much from a biblical model that we go on to identify it with the reality it seeks to portray, we may produce a form of literalism which is not merely unintelligent but comes close to bibliolatry. Possibly one reason why the Bible gives us such a wide range and multiplicity of images is that, in total, they are mutually self-correcting and self-limiting. When they are used together, they enable us to build up a balanced picture of the total reality that the Bible reveals, for the purposes which it sets before us.

There is, however, one very important difference between the models that we use in science and the models that we use in our religious thinking. In science the analogies, models and theories are all *devised* by us, even though they are guided by the data which they are used to explain. We believe that God has revealed something of his divine presence and being and that some of this has been inscripturated in the human imagery at the disposal of the biblical writers. Such biblical imagery clearly bears the imprint of both ancient near-eastern and Greek culture. It also challenges us, in our generation, to rearticulate God's revelation to us in the concepts of our contemporary culture. Without this revelatory character we would merely end up by making God in our image. For example, the biblical prohibition against making any graven image or any likeness of God (Ex. 20:4) warns us against idolatry of any kind, including idolatry in the use of any particular model which we have been given or which we have adapted for portraying God to us. We are reminded of this in those passages of the Bible which tell us that God's ways are not our ways and his thoughts are not our thoughts (Is. 55:8).

The usefulness and at the same time the limitations of models and images in both science and in religion become clearer when we compare and contrast their functions in the two fields. In science, we use a model to represent our understanding of one aspect of the physical world, and are careful always to remember that the model must be distinguished from the real world itself. In theology, it is just as important to remember that models are likewise only ways of representing God's relationship to his world and to us his creatures. The models are available in a variety of modes: in poetry, in prophecy, in chronicle and in letter; and they help us make sense of historical events from a divine point of view. They represent the relationship between ourselves and God in a variety of ways, portraying God, for instance, as father, shepherd or judge. And just as our models in science have a certain predictive value, so our models in theology produce certain expectations. For example, the model of God as father helps us to understand that God will love, care for, guide and lead us, and reveal his will to us.

In science we may hold several competing models, each of which claims to handle the same set of data. We then may apply certain procedures to enable us to judge between these models. These include such things as the parsimony of any presuppositions, the predictive range of the models, and perhaps the elegance of the particular model. The same is true to some extent in the religious sphere. If one begins from a non-theistic point of view, it may be possible to offer an alternative and non-theistic interpretation of any assertion concerning religious experience which has been made from a religious point of view. In this case we almost certainly cannot collect extra data in order to resolve the conflicts, unless we are prepared to do so on God's terms. And God's terms are that we must be willing to know and to act upon any further knowledge that he gives to us (Jn. 7:17).

A major difference between science and theology, therefore, is that in science we frequently start from raw data and a few hunches and then proceed to build up our own models to incorporate the regularities that we observe in the natural order. In theology we are guided by the contemporary pictures inscripturated by the biblical authors. These we may regard as templates against which to check any reinterpretations designed to fulfil the same function. Thus we may re-dress our religious ideas in contemporary thought-patterns, but when we have done this we must always check to ensure that the essential point made in the biblical records is still present in our revised model, and that the truth is not distorted in the process of transposition.

In science we have learned that an object in a given number of spatial dimensions cannot be fully represented in a space of fewer dimensions. Likewise in theology we find ourselves at times faced with events whose rationale is claimed to lie in eternity, that is, outside our present space-time dimensions; yet the only models available to us are those within our space-time. It is not surprising therefore that such models may appear both incomplete and paradoxical. At this point the doctrine of the person of Christ becomes crucial, for the Christian claim is that in Christ the eternal was 'projected on to' the finite dimensions of our space and time; Christ is declared to be the only complete and final projection of God the Creator into our world. In Christ we are given the fullest understanding of God and his truths that can be available to us.

Finally, we may note that in science we make use of a wide range of different models in order to do justice to the many different aspects of physical reality that we wish to study. For example, we do not falsely oppose the organismic and the molecular concepts of a cockroach, but regard them both as necessary to a full understanding of that which we are studying. So also in theology we must bear in mind that the multiplicity of images or models helps us to a full understanding of the many-dimensioned nature of divine truth, and thus circumvents the inadequacies of our earthbound thought-forms. We need all these held in a delicate balance in order to give us as full and adequate a presentation of theological truths as possible, as well as of the spiritual reality that is being communicated.

We have already suggested that the models we use in science and in theology should not be falsely opposed to one another, since they may give complementary accounts of different aspects of that which is being observed. For example, we could suggest a variety of psychological models of what happens when a person is converted, or we could present a physiological model of what happens in a person's brain at the time of conversion. And just as it is wrong falsely to oppose the psychological and physiological accounts of what is happening at conversion, so we should not set them against the account which we have traditionally given in religious terms. We want to speak of an encounter with God, of turning away from sin and idols to God, and of repentance and faith; but our account in these terms is not something which has to be *fitted in* to either the psychological or physiological accounts of the same event. Conversion makes complete sense only if we see it at all levels.

Models and myths

In the light of what we have just said, it is relevant to express the puzzlement that we find as scientists in trying to understand so-called demythologizing in contemporary theology. Some theologians argue that the essence of the gospel is obscured by the mythological language in which it has traditionally been presented; they claim that realities about God can best be dealt with without such aids, since we have now come of mature age. It seems clear to us, however, that in seeking to express the activity of God in terms which we can all understand, we must make use of images, pictures, parables and stories from the everyday happenings in the world, otherwise it is difficult to say anything meaningful at all. As scientists we have learned how necessary such models, analogies and images are in assisting our understanding of physical reality and of communicating this understanding to each other. Perhaps we may be forgiven for reminding the theologians that when we use concepts like electrons, waves, genes, species, reflex arcs, nerve nets and so on, without which we should be completely lost both in our understanding and in our communication with each other, we do not assume that the models are identical with the reality they represent. In other words, our understanding of the 'deeper truth of physical reality' makes extensive use of a wide range of concepts without identifying them with that reality. Did some of the enthusiasm for demythologizing enterprises in theology arise from a mistaken understanding of the use of models in science?

A similar point comes from the use of familiar biblical models in teaching biblical truths to young children. In teaching science we begin with the simpler and more familiar models and then progressively replace them with other models as the child's capacity for understanding grows. None of us thinks that we do any irreparable harm because we seek to match our physical models to the child's cognitive development. It seems difficult, therefore, to understand why it is dangerous (as some claim) to match our theological models to the development of a child's cognitive processes. The error that we must avoid in both science and theology is to *identify* the model with the reality which it expresses. A sure way of avoiding this is to use many different images and models to convey truth, be it physical or spiritual.

Chapter 5

The God of the physical Universe

God is the creator of the Universe. The Big Bang model is currently favoured to depict the method he used, but credible faith does not depend upon any particular model. For example, Stephen Hawking's interpretation of the initial conditions of the Universe does not affect our belief in God's creative work. The Anthropic Principle has encouraged some to restate the classical argument (for the existence of God) from design; it certainly strengthens the possibility of a viable natural theology and of a God who acts through scientific processes and not merely in causal 'gaps'. It is important to explore the different possible mechanisms, including the implications of chaos theory.

It is fundamental to biblical theology that the Universe owes its origin and continued existence to the will of God. The God revealed in Jesus is the creator and sustainer of heaven and earth. Paul is explicit: 'By him all things were created: things in heaven and on earth, visible and invisible, whether thrones or powers or rulers or authorities; all things were created by him and for him. He is before all things, and in him all things hold together' (Col. 1:16–17).

We have looked at ways in which philosophers view this. What about scientists? Can we give any credibility to a relationship between an infinite God outside time and the physical Universe? And if so, what was his method of creation and how do all things hold together?

God and the origin of the Universe

Within the scientific community, it is generally agreed on the basis of the redshifts of galaxies, the observation of the microwave background radiation, and the theoretical prediction of the observed hydrogen-to-helium ratio (and other light elements), that the origin of the Universe is well described by the model of a hot Big Bang, with the Universe expanding from a single point in time (a singularity) some 15 billion years ago. Of course, no scientific model is without its problems, and the Big Bang model as presently formulated leaves some questions unanswered, such as what came before the Big Bang and where the Universe is heading, together with technical questions concerned with the age of the Universe and the nature of the dark matter necessary for galaxy formation. Although the model has needed some careful refining, it has stood up remarkably well to thirty years of scrutiny and new observations.[1]

The model of the Big Bang has, however, stimulated a variety of theological responses, stretching back before the scientific acceptance of the Big Bang itself (*e.g.* Humphreys 1995). We need to review them as a way of getting into both the science and the theology of the origin of the Universe.

Big Bang or Big God?

There are some Christians who argue that acceptance of the Big Bang implicitly denies the existence of a creator God. They point out scientific gaps in the Big Bang model; argue for an age of the Universe in thousands rather than billions of years; assert that Scripture gives a scientific description of how God created the Universe; and therefore, believe that God is more likely to create in seven days rather than over billions of years.

Those with differing viewpoints need to be open to one another, but it must be made clear that this 'young Earth' position is a minority view, even among evangelical Christians; we are not talking here of biblical authority, but biblical interpretation. As we shall see time after time in the next few chapters, our understanding depends on whether the early sections of Genesis are taken as a scientific history of the Universe or whether their purpose is primarily theological. This is a basic question, but it is wrong to express it as a choice between alternatives. As we have seen, scientific models need to be linked together with the theological truth that the whole creation owes its existence to God. Both are needed to understand fully the origin of the Universe.

Notwithstanding, if the Big Bang is how God creates, why did he use a process which took billions of years? The scientific answer is that it takes billions of years to make carbon and other elements inside stars in order that life can exist. But why make carbon in that way? To that there is no full answer, although we must be clear that such a process enhances rather than lessens God's creative power (p. 119). Indeed, such an intricate, patient and elegant process can truly be seen as part of the 'heavens declaring the glory of God' (Ps. 19:1).

No gaps for God?

As we have seen (p. 79), Charles Coulson popularized the phrase 'God of the gaps', warning that when science has unanswered questions, we should beware of inserting God into the 'gaps' in our knowledge. As science answers more and more of the questions within its own territory, and the gaps close, God is pushed into irrelevancy. Such a danger surfaces when we face the Big Bang.

The standard model of the Big Bang raises important scientific questions as well as theological ones, and currently many of them still lack definitive answers. For example, why was the early Universe so uniform on a large scale, and why so finely tuned? When the Universe was 10^{-43} seconds old, the expansion force of the Big Bang was balanced to within 1 part in 10^{60} of the gravitational force; one second after the Big Bang, the expansion velocity was within one part in 10^{17} of what was needed for perfect tuning. That is very fine tuning! In addition, the origin of the density fluctuations which eventually led to galaxy formation is not specified. How do we explain such features of the Universe? (See Barrow & Tipler 1986.)

One possibility is that if the Universe was spatially infinite (or if there were an infinite number of universes), we could be living in a region that by chance was uniform and fine-tuned with the initial state chosen randomly. We would observe such a Universe because we are here. In a different Universe, no-one might be there to observe it. There are, however, two objections to this. First, in what sense do other universes exist if they have no observable consequences, and second, why are there so many other galaxies when only one would be sufficient for life to exist?

An alternative possibility, using so-called 'inflationary' models of the early Universe, is that quite a number of different initial configurations could have evolved to produce a Universe like the one which we observe. Such models postulate an early exponential expansion, which reasonably explains the uniformity of the Universe and its fine-tuning without assuming precise initial conditions. These models are significant, as they mean that the initial state of the Universe did not have to be chosen with great care. Notwithstanding, it cannot be the case that every initial configuration would have led to a Universe like the one we observe.

This is where the work of Stephen Hawking becomes relevant. In his best-selling *A Brief History of Time* (1988), he is particularly concerned with the initial conditions of the Big Bang. He dissents from the common assumption, made explicit by the French philosopher Laplace, that if at any one time we knew the position and speed of every particle in the Universe, we could predict its future as well as tracing its past. Hawking believes: 'it is just a pious hope that the universe is deterministic in the way that Laplace thought. The universe does not behave according to our preconceived ideas; it continues to surprise us ... The situation changed for me when I discovered that black holes aren't completely black. Quantum mechanics causes them to send out particles and radiation at a steady rate. The result came as a total surprise to me.' What this means is that information will be lost when black holes are formed and hence we can predict even less than we once thought on the basis of quantum theory. For Hawking, the Uncertainty Principle implies that space is full of tiny black holes, a hundred billion billion billion times smaller than the nucleus of the atom. And because they are so small, the rate at which information is lost is very low – which is why the laws of science appear to be deterministic to a very good approximation. 'One can calculate probabilities, but one cannot make any definite predictions. Thus the future of the universe is not completely determined by the laws of science and its present state as Laplace thought. God still has a few tricks up his sleeve' (Hawking 1988: 135).

Hawking is quite correct about the difficulty of describing the initial conditions of the Universe. Our present theory of general relativity cannot answer the question because it breaks down as one approaches the unique event (or singularity) of the Big Bang. As we go back in time, the Universe was smaller and smaller, and at the 'origin' we reach a point where there is infinite curvature and density. General relativity is unable to describe these infinites. Moreover, we would intuitively expect the effects of quantum theory to have been important when the Universe was so small. We do not yet, however, have a theory which unites general relativity and quantum theory.

Hawking aims for a complete description of the Universe in a single theory that describes how the Universe changes not only with time but also the initial conditions. At the heart of such a theory would be a unification of general relativity and quantum mechanics, in other words a quantum theory of gravity. Such a theory would answer some of the above questions by a 'theory of everything', that picks out one initial state and hence provides a coherent model of the Universe. Building on his earlier work on the evaporation of primordial black holes, Hawking believes that some of its features are fairly certain, although he acknowledges that a complete and consistent combination of quantum mechanics and gravity has not been achieved. He proposes that one way forward would be to use Euclidian space-time, which would mean that 'It is possible for space-time to be finite in extent and yet ... have no singularities at which the laws of science broke down and no edge of space-time at which one would have to appeal to God or some new law to set the boundary conditions of space-time' (1988: 136).

This leads us to another problem, because as we go back in the history of Universe, it may be that we approach, but never reach, zero time. Time apparently fades away before we reach a time of zero singularity. The Universe appears to us to have a finite past of 15 billion years, but we find no simple beginning in time with a temporal edge or boundary. It is rather like the edge of a frayed sweater. The smooth surface turns into a tangle of unravelling threads, but where does the fraying begin?

As the last of the above quotations from Stephen Hawking states, his quantum theory of gravity implies that there need be no singularity of the Big Bang at which the laws of science would break down. In real time some 15 billion years ago, the Universe would have had a minimum size corresponding to a chance quantum fluctuation from a state of absolutely nothing to a small, finite, expanding state. Put another way, quantum theory deals with events which do not have deterministic causes. By applying quantum theory to the Universe, Hawking is saying that the event that triggered the Big Bang did not need to have a cause. In this way, science is able to link the laws of cosmic change with the initial conditions from which everything began.

It needs to be stressed that Hawking's views on quantum gravity are speculative and are not widely accepted; at present it is not clear whether his ideas will work, even to the extent of whether quantum theory can legitimately be applied to the whole Universe. Some suggest that we do not know enough about either singularities or the nature of time to make a sensible judgment about them. We do not as yet have any consistent theory of quantum gravity, and even if we did, such a model may be too mathematically complicated for exact predictions to be calculated. Hawking's suggestion is one way to get around the problem of initial conditions, but the publicity it has received does not mean that it is necessarily correct (see Penrose 1989; Barrow 1991).

Hawking's model is, however, a salutary reminder to those who try to prove the existence of God by logic, that is, by arguing that as we do not yet have a full scientific description of the first 10^{-43} seconds of the Universe, there must have been someone ('God') to light the blue touch-paper of the Big Bang. But such a God would be a 'God of the gaps', implying a deistic view of God, whose only action in the Universe was to start the whole thing off. Hawking's theory of the Universe removes this particular gap, and thus totally demolishes such a false view of God. For that we can be thankful. The biblical images do not portray a God of the gaps or a deistic God, but a God who keeps the whole Universe in existence moment by moment throughout its whole history (*e.g.* Col. 1:15–17; Heb. 1:3). If Hawking is eventually shown to be correct, he is not a threat to theism; he is simply filling in another part of the 'how' story.

It must be stressed that Hawking's model does not 'disprove' God as creator, despite the assertion by some that such a scientific description leaves no room for God at all.[2] This is not the case, for at least four reasons. First, Hawking himself acknowledges the question 'Why should there be a Universe for the model to describe?' His answer is that 'you can define God to be the answer to that question' (quoted by Wilkinson & Frost 1996: 159). Even a theory of everything does not deal with the purpose or value of the Universe. Secondly, such theories are based on an assumption that we are rational beings who are free to observe the Universe as we want and to draw logical deductions from what we see. This is an aspect of the Universe which needs explanation, however, not merely assertion. Thirdly, we are still left with the question of where the laws of 'nature' themselves come from. The Universe may arise from a quantum fluctuation in a field, but where does quantum theory come from? This is not another 'God of the gaps' idea, because we are talking not about a gap in scientific explanation but about the origin of science itself. Finally, the claim that it disproves the need for a creator is really a 'conflict' approach to the relationship between scientific and theological description, arguing that both descriptions say the same kinds of thing about the same event: in other words, that the existence of a scientific description of an event automatically invalidates its consideration as God's creative activity. This assumption disallows the possibility of complementarity between scientific and religious descriptions; if we do not make the assumption, we can have a totally naturalistic account of creation without ruling out a providential account.

A surer path to God?

With the Copernican revolution, humankind was no longer perceived to be at the centre of the Universe. This change is summarized in what has come to be known as the Cosmological Principle: our position in the Universe is in no sense preferred. Many scientists, however, are now accepting anthropic reasoning, in which humankind again plays a key role in understanding our Universe. For example, Barrow & Tipler (1986) point to the age and size of the observable Universe, nuclear energy levels in beryllium and carbon, and the remarkable properties of oxygen in the atmosphere. All these things are essential to the existence of carbon-based life, and are determined by subtle balances among the forces of nature.

It is these aspects of the Universe that have led to an interest in the 'God question' among many scientists. Neither Paul Davies, Professor of Theoretical Physics at Adelaide, nor Sir Fred Hoyle has any time for traditional Christian theism, but both are among recent writers who cannot leave God alone. Davies even claims that 'in my opinion science offers a surer path to God than religion' (1983: ix).

Davies is particularly impressed by the fine-tuning of the physical constants within the Universe, and sees this as the most compelling evidence for an element of cosmic design. He believes that there is no conflict in the idea of a Universe evolving according to the laws of physics but nevertheless subject to intelligent control. From this basis, he suggests a 'natural God' operating within the laws of nature, directing and controlling the evolution of the cosmos; in other words, an intelligence firmly within the Universe and subservient to it. The development of humankind's future control over the Universe becomes a model for a controlling intelligence within the laws of physics. Hoyle (1983) has argued along similar lines.

Why are such people led to make such claims? Some have suggested that it is in response to a basic need to feel that there is purpose at work and hope in the world. This is probably part of the answer, but another likely factor is that they are a response to a sense of 'awe' at the Universe. Some of the anthropic balances in the laws of physics are indeed awe-inspiring and prompt the question 'Why does this happen?' This sense of awe is shared by many scientists who have no 'religious' convictions.

Hoyle has always been quite clear about his negative views on Christianity, although he does recognize some sort of religious impulse, a feeling of being derived from 'something out there'. In contrast, Paul Davies explicitly rules out any attempt to discuss religious experience, revelation in the Bible or morality. This unwillingness to take account of these things is one of the reasons why both Hoyle and Davies posit a God who is not transcendentally other but merely a 'demiurge'. They come to this position by putting together the apparent design of the Universe and their belief in future human progress. Their god is not transcendent, but contained by the Universe and its laws. Neglecting revelation and religious experience, it is not surprising that Davies and Hoyle illustrate the criticism of Hume and Kant that the design argument leads not to an infinite creator but, at most, to a cosmic architect using existing material (see p. 36).

Such suggestions about design have to be seen against debates on the Anthropic Principle, which is a major theme in modern cosmology. It is most commonly formulated in either a weak or a strong form, first distinguished by Brandon Carter (1974).

The Weak Anthropic Principle (WAP) states:

> Observed values of all physical and cosmological quantities are not equally probable because we must take into account the fact that our location is necessarily privileged to the extent of being compatible with observers.

The Strong Anthropic Principle (SAP) is:

The Universe must be such as to admit the creation of observers within it at some stage.

It should be recognized that the WAP is not suggesting that humankind is the cause of any of the physical properties of the Universe, but that our presence acts as an observational selection effect on the kind of Universe that we find ourselves in. As a trivial application of weak anthropic reasoning, we can note that it is no surprise that we do not find ourselves living on the surface of the Sun or in outer space, but rather that our location is correlated with a planet conducive to life. Similarly, it is uncontroversial to explain the large size and old age (on human scales) of our Universe by reference to the fact that the Universe has to be old enough for the atomic elements necessary for life to have been created by stellar processes and then for life to have emerged and evolved up to the level of humankind. When applied to the cosmic coincidences, however, the WAP has no force unless it has an ensemble of universes on which it can act as a selection effect.

The Strong Anthropic Principle is much more controversial, and is really a quasi-religious statement. The main attempt to justify it scientifically makes use of the role of the observer in quantum theory: the fact that an observer is necessary for the collapse of a quantum mechanical wave function suggests that only universes with observers are actualized. This interpretation of the SAP has sometimes been referred to as the Participatory Anthropic Principle. Most, however, find very implausible the idea that an observer who appeared billions of years after the beginning of the Universe could have brought it into being.

The main alternative to the Anthropic Principle simply invokes a designer as the cause of the Universe's life-permitting properties. As even some agnostic scientists are prepared to admit, in many ways this 'design principle' is the most obvious explanation. For example, Paul Davies concludes that 'These rules [laws and initial conditions] look as if they have been designed. I do not see how that can be denied' (1992: 214).

The Christian response is more positive, and stresses that although something of God can be seen in his creation, it is only his self-revelation that truly reveals his character. That this God who is greater than the Universe has revealed himself within the Universe means that we can know of his existence and nature. Neglecting this revelation leads to the danger of pantheism, with the distinction between nature and divinity becoming blurred; the 'natural God' of Davies could equally be 'Nature'.

Fruitful interaction

Is there a way to avoid the 'God of the gaps' while taking seriously the way the Universe is? Indeed there is, and it rests on a number of theological and scientific foundations.[3] It is quite possible to agree with Hawking and at the same time to advocate a scientific explanation of the origin of the Universe without invoking a God of the gaps. It is important to recognize, however, that such an explanation is not complete in itself.

Accepting the fact that God has revealed himself in the Bible, one is led to the conclusion that the Universe is neither self-generating nor self-sustaining, but has the status of creation; the character of God's creative activity is as originator, sustainer, governor and provider of that creation. Due to the time and the manner in which the Bible was written, divine action is described in figurative, anthropomorphic and poetic terms rather than scientific language; Genesis necessarily gives a poetic or liturgical story of origins (Blocher 1984). The revelation of God in the Bible establishes him as Creator and sets out the status of creation, but to obtain specific details of cosmic history one must turn to an observational study of creation itself. Cosmology does not give us the purpose or value of the Universe; the scientific 'how' needs the 'why' of the biblical account.

This leads to the unavoidable implication that the world is not neutral but must bear marks of the Creator's character, however veiled. Some see these in the Anthropic Principle, but, in a more general sense, such a belief is the basis of a new interest in natural theology among scientists who stand directly within the Christian tradition. They see the discoveries of modern cosmology to be of the utmost importance for their view not only of creation but also of the Creator. For example, John Polkinghorne (1986; 1988) has argued that science and theology should be seen as capable of fruitful interaction, and detects this fruitfulness in a revival of natural theology. This is not a resurrection of proofs for the existence of God, which was the traditional context of natural theology, but a recognition of possible insights into the way the world is.

Therefore, instead of trying to prove God by the classical cosmological argument that there must be a first cause to the Universe, science discloses the intelligibility of the Universe. It is a striking and non-trivial fact that the pattern of mathematics is realized in the physical structure of the world and that our minds are able to solve problems that the physical world presents. In this recognition of intelligibility, Polkinghorne sees the most reasonable explanation to be the existence of a creator who is the common ground of the rationality of our minds and of the Universe.

Likewise, rather than trying to prove the existence of a designer, examination of science discloses a Universe of anthropic fruitfulness, finely balanced for life to exist. Science may be successful in explaining how some of these balances come about, but its models are themselves possible only because of the nature of fundamental laws. An explanation of economy and elegance would be that the world is created by a creator who wills it to have this anthropic fruitfulness.

This revised and renewed natural theology points to the scientific givenness of law and circumstance rather than to particular occurrences; it is not a return to a God of the gaps, but a response to the world that science discloses. It does have its limits, however, bringing us only to a 'cosmic architect' or 'great mathematician' rather than to a full knowledge of the God of Christian theism. For the latter, revelation is needed.

This position is consistent with the wisdom literature and other parts of the Bible which encourage Christian thinkers to seek evidence of God in the Universe as well as in the Scriptures (*e.g.* Ps. 19:1; Acts 14:7; 17:22–31; Rom. 1:19ff.). Such a natural theology, however, needs to be brought within the embrace of revealed theology. It is primarily because Paul Davies neglects experience and revelation that he strays into pantheism and narrows his approach to exclude any personal knowledge and understanding of God. Within the context of a God who reveals himself, one is able to hold together immanence and transcendence, human fallibility and human responsiveness, the insights of creation which point to a creator and those factors which do not. It is significant that the biblical passages which seem to encourage natural theology stress that this general knowledge of God needs correcting and expanding by special revelation (Kidner comments on the research scientists' mandate in Ps. 111:2, 'Great are the works of the LORD, studied by all who have pleasure in them', that 'verse 10 must be its partner, "The fear of the LORD is the beginning of wisdom; a good understanding have all those who practise it", lest professing to be wise we become fools like those in Romans 1:18–23': Kidner 1975: 397, quoting from the Revised Standard Version).

God's action in the world

Is it right to pray for rain? The motivation for such a prayer may be understandable, but does it make theological sense? Can God so work in the world that those with a scientific worldview can accept the physical laws which produce rain while at the same time allow God to have some sort of control that can be seen as the answer to prayer? And if God can make it rain, can he change the seasons, making summer occur before spring?

These are not meaningless questions. Neither are they new ones. Questions concerning how God works in the world have a very long history in Christian theology, but the emergence of modern science has sharpened them in both the theological and popular arenas. Traditional Christian theology understood God to have a personal and particular concern for the unfolding histories of his creatures. His action in the world was seen not only in creation, but also in the exodus and in the cross. God is at work in the world, but how does he do it?

We have already seen that God upholds our world from within, rather than interfering with its operation from without (p. 40). Our understanding leads us to two immediate problems. First, how can a personal God act as an agent within the regularly ordered world revealed by scientific study? The very laws of nature, taken by some to testify in their rational beauty to God's existence, apparently leave no room for his free action. The argument here is that if God creates a Universe governed by faithful laws which allow us to understand its working and predict its future, however imperfectly, then it is difficult to see how God may act without breaking his own created causal network. Rain is governed by the weather system which ultimately depends on the physics of the atmosphere. How can God cause it to rain without breaking those physical laws?

Secondly, what about evil? If God does act in a particular way, why does he not do it more often? If he heals someone of back pain in a healing service, why does he seem to stay inactive in the face of earthquakes and the Holocaust? In fact, the more strongly one speaks of God's particular action in the world the more forceful becomes the problem of evil. Some modern theologians who show reserve about a God of particular providential acts seem to do so in order to give a defence to the problem of a natural evil in terms of the regular structure given to the world. That is, if God allows the natural world to proceed without intervention, then he must allow the good and the bad.

In recent years a number of approaches have attempted to overcome these problems. Some have tried to maintain the traditional understanding of God's action, while others have been forced radically to reformulate it.

Models of God's action in the Universe
The 'all in the mind' God

The existentialist approach of Bultmann and others draws a distinction between the 'exterior' world of science and the 'interior' world of religion, making a fundamental separation between our knowledge of physical events and the God who is known in experience. This interpretation says that God does not act in the physical world in any particular physical way, but achieves his purposes by 'acting' in the person of faith as he or she encounters God's Word. Prayer for the end of a drought will not lead to God making it rain, but to the praying person being moved to help in the midst of the drought.

In practice, this view depends on belief in an old-fashioned Newtonian mechanistic world which was totally predictable. Modern science is more subtle than that. An additional problem is that it assumes a distinction between religious and scientific knowledge, and is open to the obvious question, 'How does God do this?' We may not understand the exact relationship between mind and brain, but we do not doubt that there is a relationship. How does God change a person's mind without in some way having to interact with the physical neural network? It is simplistic to divorce the mind from the physical. Even a model of God changing a person's mind implies some particular interaction of God with the physical world (see p. 181).

The 'sit back and watch' God

Oxford theologian Maurice Wiles's book *God's Action in the World* (1986) was caricatured, perhaps unfairly, as 'God's Inaction in the World' (Hebblethwaite 1989). Wiles argued that God is creator and sustainer, but that his action is limited to the one great act which caused and keeps the Universe in being, allowing radical freedom to human creatures but requiring radical self-limitation on God's part.

Wiles imagined a play in which the author provides the characters and setting, and the major plot follows the intention of the author; but the actors have the freedom to determine their own fate. Evil thus becomes a risk taken by God in allowing freedom to live within a physical-lawful environment. For Wiles, providence becomes a kind of teleological insight into the general physical process, with religious experience recognizable only in retrospect. Conversion on this interpretation is not God forming a particular relationship with an individual but simply an acknowledgment that God was active in bringing the world into being.

This is almost a trivialized view of God. Certainly, part of the expression of God's work in the world will be the reliability and beauty of the laws of nature; but in addition we must agree that God has a consistent rather than a fitful relationship with his creation.

Wiles's approach is little more than a sophisticated version of deism, in the sense that God is seen to start the Universe off and then have nothing more to do with it, particularly in specific acts. The Universe becomes a 'no-go' area for God to work, and consequently it is difficult to see how God can be spoken of in terms of personal relationships. We might also ask, 'If God allows freedom within the Universe, why does he not allow himself a degree of freedom to work within that Universe?' To deny God total freedom does not necessarily mean that he has no freedom at all.

Furthermore, is it possible to reconcile this view with the historical view of the incarnation presented in the New Testament? The resurrection stands as firm evidence that God has intervened in specific ways in human history and therefore in the physical Universe.

A 'persuasive' God

An extension of the 'all in the mind God' and the 'sit back and watch God' uses an analogy between God's action and our experience as agents, and from this attempts to assimilate the nature of the Universe to our human nature. This is a version of process theology or 'panpsychism' which sees each event in the Universe as the selection of possible outcomes, followed by actual realization. Each event has a psychic pole and a material pole; God works as an agent at the subjective stage, exercising his power by persuasion or lure rather than coercion.

The attraction of this is that God is able to involve the physical while interacting with the 'spiritual'. There is, however, no evidence whatsoever that the physical world has such a nature, and it is not easy to see how the psychic and material poles are connected. Secondly, and worse, if carried to a logical conclusion, it means that primitive objects such as quarks have an ability to 'select' outcomes, which seems ludicrous. Thirdly, it is difficult to see how God can do anything of importance at such a level. In terms of physical interaction, it differs little from Wiles's approach. God becomes a passive deity, and in Mascall's words, 'more to be pitied than to be worshipped'.

A 'bodily' God

Process theology uses an analogy between God's action and our action. It implies 'panentheism', which extends the analogy, attempting to assimilate God's action in the world to our action in our bodies. This is attractive to feminists. For example, Grace Jantzen (1984) sees the world as God's body, God working in it just as the soul works within the body. Arthur Peacocke (1979) goes down a similar path, though not to the same extreme, by seeing the Universe as like a foetus in the womb.

Such an approach can sound superficially attractive, but does not stand close examination. First, the analogy suffers from the fact that it is difficult to understand what such embodiment means. How are mind and brain related, and what does it actually mean to say that 'the soul works in the body'? Secondly, if the Universe is in some way God's body, then God becomes totally vulnerable as the Universe changes with time. The analogy may be adequate for a 15-billion-year-old creation when the Universe has order and discernible structure, but is totally inappropriate when the Universe was a quark soup. And what was God like before the Big Bang? Thirdly, such an analogy sees the nature of the physical world as an organism having unity in its overall structure. Although favoured by some who have tried to use modern physics to defend an eastern view of the world, the Universe is just too subtle to fit the picture. In some senses it shows 'organism' qualities, in other senses 'mechanistic' ones (of which we say more below). The Universe is not a body; it is a complex admixture of many things. Fourthly, panentheism threatens God's otherness and freedom, and hence compromises the world's freedom to be itself.

A 'chaotic' God

John Polkinghorne (1989) has suggested a new way of looking at providence. His starting-point is that if there is room in the physical world for our own exercise of free will, then surely God must enjoy a similar room for manoeuvre. Polkinghorne is not the first to propose this idea, but he extends it to locate that room for manoeuvre within chaotic physical systems.

For the past few decades there has been a growing realization that physical laws do not provide an exhaustive description of the world, nor do they imply inflexibility of action within the process they describe. Quantum theory is an illustration of such a point, indicating that at the atomic level there is a basic uncertainty which is independent of how good our measuring apparatus is. Pollard (1958) argued that this uncertainty of quantum events may be the site of God's free and 'cloaked' actions in the world. Events at the atomic level, however, are not always the same as at the scale we work on as human beings. The atoms in my body have a non-zero probability of passing through the atoms of a wall. But this does not mean in everyday life that there is no point in using the door!

Polkinghorne suggests that it is in chaotic systems that God has freedom, and that his action is unable to be directly seen. Chaotic systems have a great advantage over quantum systems in that their effects are felt at the scale of everyday life.

What are 'chaotic systems' in this sense? They were discovered in 1961 by Ed Lorenz, a meteorology professor at the Massachusetts Institute of Technology. He was interested in air movements in the atmosphere and tried to model them. He would type in the starting conditions on his computer, which would solve the equations and give a prediction. In one of his computer runs he typed in a number which was incorrect by one part in a thousand. His common sense told him that this small error would only cause a very small difference in the final prediction. Much to his surprise, he found that this small error in the starting conditions changed the prediction enormously. What he had come across was a chaotic system.

Such systems exhibit a great sensitivity to initial conditions, with very different outcomes arising from infinitesimally different starting-points. Lorenz named the phenomenon 'the butterfly effect', picturing the weather system as so sensitive that (in his words) the flapping of a butterfly's wing in Rio could lead to a hurricane in New York.

The weather is a good example of a chaotic system. The physical laws which determine the weather are well known, and forecasters use them to make weather predictions. Why then do they get things wrong, and why do forecasts get more imprecise the further ahead they predict? Chaos explains this by saying that even if the laws are known, the precise predictions are extremely sensitive to the initial conditions. However assiduously the Meteorological Office collects data on pressure and temperature on which to base their predictions, there will always be a degree of uncertainty. This uncertainty is amplified in a chaotic system. This is not to say that gross features like global warming cannot be predicted. It is just to say that precise predictions are impossible.

The importance of chaos is that, in contrast to the apparently 'clockwork world' of Newtonian mechanics, there are systems obeying immutable and precise laws which do not act in predictable and regular ways. When the dynamics of a system is chaotic its outcome can be predicted only if the initial conditions are known with absolute precision. This is obviously impossible; even a computer as large as the Universe would not give such precision. This means that for finite beings there is an inevitable uncertainty within the everyday world even if the laws of physics are completely known. It must be stressed that not all systems are significantly chaotic, and for such systems the Newtonian picture is still valid. For example, the planets orbit the Sun in a regular and predictable way, at least to the accuracy required in practice.

The sensitivity of complex systems to initial circumstances means that they are intrinsically unpredictable. Polkinghorne then takes a significant step. Viewing science as a critical realist activity, he wants to make an additional inference of a strong link between epistemology and ontology; in other words, to bring together what we know of the world and what the world is truly like. On this basis he argues that chaos means that there are systems in the Universe which are inherently open to the future; they are unpredictable and undetermined.

From this, three things follow. The first is that here is a genuine ground for human free will. Secondly, the future is not implied by the present, and thus any kind of Laplacian determinism (p. 91) is ruled out. Thirdly, Polkinghorne argues that God is at work in the flexibility of these open systems as well as being the ground of law; God's particular activity is real although it is hidden from us.

This is theologically innovative and interesting. Providence becomes a subtle interaction between our personal freedom, the freedom inherent in the physical nature of the Universe, and God's freedom. On this view, God does not know the future because the future is not yet determined. Miracles, being events of radically unexpected character, are seen as an outworking of this openness in particular circumstances. Natural evil thus has a free-process defence analogous to the free-will defence for moral evil, in that the openness that the Universe has in exploring its potential can sometimes be for good and sometimes for evil. Intercessory prayer is defended in its traditional sense. Therefore, is it right to pray for rain? Yes it is, according to Polkinghorne, because the weather system is a chaotic system showing openness to the future. Is it right to pray for summer to come before spring? The answer is no, for the seasons are determined by the simple and essentially non-chaotic system of the Earth's rotation about the Sun.

What do we say about such an interpretation? It certainly takes modern science seriously in a way that many of the other approaches do not. Chaos is a major discovery of our time and reveals that our mechanistic view of the world is only a small part of the story. Furthermore, Polkinghorne defends God's freedom to work in acts of special providence while self-limiting the areas of those acts.

There is reason for caution, however. First, it is not clear that an epistemological openness need imply an ontological openness. Chaotic systems cannot be predicted by finite beings. But that does not stop an infinite being theoretically having the ability to predict the system (Stewart 1989). Secondly, the charge of a 'God of the gaps' may be levelled at Polkinghorne's argument, albeit in a different sense from that originally envisaged by Coulson. You can understand a chaotic system in terms of its laws, but you cannot predict its future. God may be a 'God of the gaps' in the sense of being confined to work in the gaps of scientific prediction. But is God so confined? Has he so limited his activity that he works only in chaotic systems and in a way that is hidden? The Bible is clear that God's providential work in history is seen, recorded and communicated as a basis of faith.

A 'big enough' God

Austin Farrer (1967) suggested that God's action in the Universe should be described in terms of 'double agency'. He argued that it is impossible to conceive of God's ways of acting in terms of our own, and therefore the causal joint between God's action and ours will always be hidden. Consequently each event in the Universe must have a double description, and can therefore be spoken of in terms of the providential action of God while at the same time having a full natural explanation.

Such a view laid the foundation for much evangelical thinking. As we have seen (p. 38), the late Donald MacKay vigorously defended a strong view of God's providential guiding of history while at the same time allowing for a complementary description in terms of natural processes. Moreover, God is outside time as well as outside space. This is more difficult for us to imagine. Perhaps the nearest we come to experiencing timelessness is in those comparatively rare flashes of inspiration when, for instance, the solution to a complex problem we have been thinking about for weeks comes to us in a moment. Although not apparently going through the sequence of argument, we feel convinced in that moment that we have hit on the solution.

Roger Penrose (1989: 541–547) cites some examples of such moments of inspiration experienced by mathematicians, pointing out that it is often the apparently aesthetic character of the solution, its beauty or its elegance, which provides the inspirational appeal of the solution-in-a-flash. Artists can experience similar inspiration. A striking example is provided by Mozart's apparent ability to experience and appreciate, at the moment of composition, a lengthy piece of music all at once: 'My mind seizes as a glance of my eye a beautiful picture or a handsome youth. It does not come to me successively, with various parts worked out in detail, as they will later on, but it is in its entirety that my imagination lets me hear it' (Hadamard 1945: 16).

To picture God outside time is not to imagine him as static or uninvolved but as seeing creation – its complete span of space and time – as a whole. The purpose-making, the planning, the unfolding of the drama with all its interconnected parts, combine to make up that whole.

We may find it just about possible to conceive God within time and even of God outside time, but thinking of God as both together really is difficult. John Polkinghorne (1989: chapter 7) talks of God 'being' (God outside time) and God 'becoming' (God within time) as the two opposite poles of the model, and Donald MacKay (1988: 193) distinguishes between two persons in one Godhead: God-in-eternity and God-in-time; God transcendent and God immanent.

MacKay's thesis differs fundamentally from that of Polkinghorne. MacKay believed that God's sovereignty is unaffected by the discoveries of modern science. No hidden gap is needed, for God is able to write into the history of the Universe whatever he chooses. This is backed up by what MacKay called 'logical indeterminacy', the idea that no human agent can give an absolute prediction of the future without itself changing the conditions on which the prediction was made (see pp. 41, 181). This means that the future cannot be determined, even in a totally deterministic world.

Conclusions

There is no easy answer about how God relates to the physical Universe, and there are many questions for which we do not have answers. We do not know if the Earth will become more and more inimical to life, or whether the end will be sudden, so that 'the day of the Lord will come . . . [and] the heavens will disappear with a roar; the elements will be destroyed by fire, and the earth and everything in it will be laid bare' (2 Pet. 3:10–11), or even whether cosmic dissolution will mark the Lord's coming. We do not know if there is intelligent life elsewhere in the Universe. But from the practical point of view, these questions do not matter. We have no hesitation in asserting our firm belief that everything in the Universe is kept in existence by God's sustaining activity, and that the laws of science are a reflection of that faithful activity.

In the ongoing discussion a number of points must be kept in mind.

First, we need to be very careful in understanding the nature of science and its current position. Those who have ruled out God's particular actions in the Universe tend to do so on the basis of an outdated, mechanistic view of the laws of physics. Science can never give us an exhaustive description of the world, and scientists such as MacKay, Houghton and Polkinghorne are quite right to be open to unexpected phenomena in the Universe. Theology must be serious in its interaction with science but must not be dominated by it. Whether or not 'chaos' is an avenue of God's action in the world, it helpfully reminds us that there is more to the Universe than we often think, and that physical systems are less easy to describe fully than we once thought.

Secondly, we need to affirm the tension between the biblical teaching on human freedom and on divine sovereignty. Evangelicals have taken different views on how to affirm these, and both Arminian and Calvinist approaches recur in modern writings; human freedom is required for human responsibility to be real, but at the same time God remains in control. This persisting tension is a reminder that any one view of providence might be neat and simple in a philosophy textbook but may be far too simplistic to do justice to a complex Universe and the God who is beyond that Universe. It is easy to espouse a simple philosophical or theological system and ignore the biblical data.

Thirdly, we need to explore the biblical tension between law and grace in terms of God's action in the world. God is a moment-by-moment sustainer of the physical laws, not a mere remote ruler. This means that science is simply describing his normal mode of working, a way of 'thinking God's thoughts after him', as Kepler put it. On this basis, order in the Universe is a reflection of God's faithfulness in creation. God has a consistent relationship with the Universe which allows us to do science, and more importantly, to learn in a reliable environment. In stressing that truth, however, we shall be wrong if we devise a way of thinking about God which limits his freedom. Miracles can be seen as special acts of grace when God supersedes his normal ways of working. But we must beware of distinguishing too sharply between different workings of God. Too many miracles would deny the Universe its basic order, negating the possibility of responsible learning; too much law denies God any freedom at all within his created Universe. It may be that we need a more personal rather than scientific analogy for God's action, emphasizing the tension or even mystery in God's maintaining an ordered Universe while working in unusual ways for specific purposes.

Chapter 6

Creation

Understanding of God's creative activity involves interpreting Bible passages such as Genesis 1–3. God's creation proceeds from chaos through massive geological and biological changes to humankind. The essential distinguishing trait of humankind is God's image, not an anatomical or genetical trait; the act of disobedience which separated humans from God, otherwise known as the fall (Gn. 3), had the consequence of spiritual death. Women and men are complementary to each other. Human behaviour manifests freedom to choose; we are not determined by evolutionary history, genetic predisposition, or psychological processes. Attempts to base ethics on the direction or results of evolution fall into the naturalistic fallacy (i.e. the assumption that what is also ought to be).

There are a number of creation accounts in the Bible, but the most detailed and debated ones are those in the first two chapters of Genesis.

Biblical interpretation

The creation stories in Genesis 1 and 2 are utterly different from any description that would appear in a scientific work, and, as we have repeatedly emphasized, it would be wrong to read them as if they were scientific accounts. Arguing from the text itself rather than from any external factors which might be pertinent, Henri Blocher points out that Genesis 1 appears to be a celebration of God's creation as a work of six days followed by a day of sacred rest, and the latter is undoubtedly intended to be an allusion to the Sabbath (Ex. 16:29; 20:8–11; *etc.*). 'How are we to compare the assertion of the seven days with the billions of years, at the lowest estimate, which current scientific theory attributes to the origin of the universe?' (Blocher 1984: 40).

There are at least five ways of interpreting the 'days' of Genesis 1.[1] A key to understanding this passage is the meaning it would have had for the original writers and readers. Study of the ideas of creation in the Near East in Old Testament times throws much light on Genesis 1.[2] The more closely one looks at Genesis 1 in the light of the religious ideas with which the Hebrews had to do battle, the clearer it is that the meaning of the passage is essentially theological, not historical or scientific. It deals with the questions which theology asks, not those which science asks (Lucas 1989: 93).[3]

The question remains, of course, whether the creation of the world was *in fact* sudden or gradual. But we are in error if we insist *in principle* that creation in six days of twenty-four hours is any more consistent with God's nature and his supernatural power than creation over a long space of time (p. 119). The word *bārā'* (create) is used in Isaiah 43:1, 15 of the creation of Israel, a process spread over several hundred years from Abraham to the covenant-making at Sinai. The conclusion from a detailed inspection of the Genesis 'days' is that Genesis 1 does not state or deny either a short or extended span of time. It is only when we look at such facts as we can discover from examining the world that we have to accept that things are not as they were when the world was first created: in the past there were very large reptiles (dinosaurs), and now there are none; there used to be vast forests made up of kinds of trees and ferns that do not now exist (except fossilized as coal beds); new volcanic islands have appeared; river valleys have been deepened; and so on. The changes are due to processes of some kind or another, even if they operated faster in earlier times than now, and even if they involved catastrophic events like a great flood. In other words, it is consonant with the nature of God as revealed throughout Scripture that in 'sustaining all things by his powerful word' (Heb. 1:3) he unfolds his purposes through natural processes as the world reaches its present state. We must retain open minds about the speed and time of his methods if we are to be faithful to Scripture.

The main discontinuities in the Genesis 1 story are the divisions between the days; if the passage was not expressed in this way, it could be read as a continuous creative progression from disorganized inorganic compounds to humans made in the image of God. Ascertaining the correct meaning of the 'days' is far from simple. But there is another ground for arguing for an unevenness in the progression of creation, and that is the special involvement of God in the creation of matter and of humankind. The word used of God's activity in these contexts is *bārā'* (create; 1:1, 27 [three times]; 2:3, 4). On other occasions (except in 1:21 – see p. 264), the word *'āśâ* (make or mould) is used. *Bārā'* emphasizes a special divine action, in contrast to *'āśâ*, which has more the sense of 'modelling' pre-existing material (like a potter making a pot). All creation is a divine activity, but this passage could be interpreted to mean that some events may have been some sort of 'special creation' rather than 'divine creation through natural processes' (see below).

Humankind and imago Dei

So far we have referred to the human race only in passing, but in many ways the nature of humankind is central to Christian debates about creation. Put crudely, are apes 'on the way up', or are they distinguished from the rest of the animal creation by divine intention?[4]

The key factor in understanding our nature as taught in the Bible is not any physical or functional trait, but the 'image of God'. This is how the Bible distinguishes us from the other animals. 'God created man in his own image' (Gn. 1:27).

Because Christ was God, and took upon himself the form of a human being, it is easy to fall into the trap of assuming that God's image is the same thing as our human form. A moment's reflection shows the naïveté of this; we have already seen that the Bible refers to God as having anatomical parts (eye, arm, *etc.*; even wings, *e.g.* Ps. 36:7), but only when it describes an activity or function (p. 84). The idea of God's image in us being the same as our physical body is as far-fetched as the scholastic belief that God resided in the pineal gland (largely on the basis that no function was known for it in medieval times).

At one time, theologians tended to equate God's image in humankind with rationality. There is no scriptural support for this, and studies on animals show plenty of evidence for rudimentary thinking and learning.[5]

Modern theological opinion is united in agreeing that the *imago Dei* is non-anatomical; theologians regard God's image as a relational, not a physical, entity. For example, Emil Brunner, commenting on 2 Corinthians 3:18, noted that 'man's meaning and his intrinsic worth do not reside in himself but in the One who stands over against him . . . Man's distinctiveness is not based upon the power of his muscles or the acuteness of his sense-organs, but upon the fact that he participates in the life of God, God's thought, and God's will, through the word of God' (Brunner 1939: 96). C. F. D. Moule (1964) concluded: 'The most satisfying of the many interpretations, both ancient and modern, of the image of God in man is that which sees it as basically responsibility (Ecclesiasticus 17:1–4)'. H. D. MacDonald has proposed that 'image should be taken as indicating "sonship", which holds together both the ontological and relational aspects of the image' (1981: 16).

Twentieth-century neo-orthodoxy has understood 'image' mainly in terms of relationship. Another facet is 'representation'. The original word suggests a statue or something similar. Scholars have pointed out that in the ancient Near East, when a king conquered territory he would set up an image of himself there to express his claim of dominion over it in his absence. Moreover, the king was often thought of as the representative or image of the national god, sometimes being regarded as the god's adopted son. Applying this to Genesis 1:26–28 leads to the following conclusions.

1. It is the *whole being*, not some element in the human constitution or some attribute, which constitutes the image of God.

2. As those who image God to the world, humans are God's viceroys, given dominion over the earth and its creatures.

3. This involves a relationship of responsibility to God because the viceroy is answerable to the absolute ruler from whom authority is derived. This remains after the fall, but in the distorted form of a relationship of rebellion.

4. All men and women[6] (not only kings) are created in the image of God.

5. We are created 'as' the image of God rather than 'in' that image.

Our humanness is not located in any specific physical part of us. Although our spiritual nature is physically embodied, it is not the same thing (1 Cor. 15:42ff.), nor is it correct to regard humans as animals with an added soul (p. 140). We return to the nature of humanness later (pp. 204ff.).[7]

Once we accept that our spiritual nature is not the same thing as our bodily envelope, our physical ancestry and genetic relationships fall into perspective. *Homo sapiens* has one of the best fossil records of any animal species, despite the criticisms that can legitimately be made of the over-enthusiasm of palaeoanthropologists. Although anthropologists argue about the details of the relationships between apes and possible human ancestors, there is no major dissension about the general outline of primate history (see, *e.g.*, Jones, Martin & Pilbeam 1992).

Notwithstanding our close genetic relationships to other animals, these studies are not relevant to our understanding of man if our distinctiveness as humans is spiritual rather than physical. We are apes, but distinctly different, with a unique capacity for relationships. The key problems concern our moral nature and relationship to God, not our physical ancestry. As B. B. Warfield (1915) commented, 'it is to theology, as such, a matter of entire indifference how long man has existed on earth'.

The fall

If Darwinian humankind is 'evolving upwards' from the apes, they might be expected to be getting better all the time, morally as well as physically (p. 235). This contrasts with the scriptural position that we are a special creation, inbreathed by God at a specific point in time. Was there a single historical Adam and a single historical Eve? Paul's teaching seems to demand this (Rom. 5:12, 17; 1 Cor. 15:21, 45; but see p. 115). This view may be related to Paul's teaching that 'sin entered the world through one man, and death through sin' (Rom. 5:12). Three points need to be made.

First, the death that came into the world was spiritual (separation from God), not physical. Adam and Eve 'died' the day they disobeyed (Gn. 2:17), but they survived physically (and produced all their family) after their exclusion from God's presence. The cycle of (physical) death and new (physical) life was established in the natural order; the Bible record is concerned with the rift introduced into this order by human disobedience, that is, spiritual death. As John Stott wrote in his exposition of Romans, 'Death is represented in Scripture more in legal than in physical terms' (1994: 171). The curses in Genesis 3 are primarily concerned with life, not death; they affirm that hardship and wretchedness will continue until death supervenes. Ernest Lucas puts it in these terms: 'Death [in the OT] is seen as a spiritual power, not just an event at the end of life. It is a power which weakens and diminishes life, eventually leading to its end. One falls into the power of death when cut off from God as the source of life (see Dt. 30:15–20)' (Lucas 1989: 110). Christ's death on the cross was to 'bring us to life in him' (reconciliation and spiritual life), not in the first place to defeat or overcome physical death.

Secondly, Adam is described in the Bible as a farmer; his son Cain became a city-dweller; while Adam was apparently still alive, his descendent Tubal-Cain was born, 'who forged all kinds of tools out of bronze and iron' (Gn. 4:22, 25). This places Adam firmly at the end of the Neolithic period, around 10,000 years ago. But by the time Neolithic farming was beginning in the Middle East, *Homo sapiens* had spread to many parts of the world: there were Indians in America, Aborigines in Australia, and so on. A Neolithic Adam and Eve could not be the physical ancestors to the whole human species. But we have already seen that the key to understanding our nature is that we are made in God's image; our physical nature is secondary to our status as '*Homo divinus*'. Spiritual inbreathing and spiritual death are not determined by or spread through Mendelian genes: they depend upon God's distinctive methods of transmission. As Kidner has pointed out:

> With one possible exception[8] the unity of mankind 'in Adam', and our common status as sinners through his offence, are expressed in Scripture not in terms of heredity (Is. 43:27) but simply of solidarity. We nowhere find applied to us any argument from physical descent such as that of Hebrews 7:9, 10 (where Levi shares in Abraham's act through being 'still in the loins of his ancestor'). Rather, Adam's sin is shown to have implicated all men because he was the federal head of humanity, somewhat as in Christ's death 'one died for all, therefore all

died' (2 Cor. 5:14) . . . After the special creation of the first human pair clinches the fact that there is no natural bridge from animal to man, God may have now conferred his image on Adam's collaterals to bring them into the same realm of being. Adam's 'federal' headship of humanity extended, if that was the case, outwards to his contemporaries as well as onwards to his offspring, and his disobedience disinherited both alike (Kidner 1967: 73).

The Bible insists on the spiritual unity of the human race (Acts 17:26; Rom. 5:12–14). This does not necessarily mean a genetical unity.

Thirdly, the New Testament passage which most explicitly refers to the fall is Romans 8:18–23. This clearly teaches that the whole of creation (including humanity) has been affected by the presence of sin in the world. A closer examination, however, shows that the fall primarily involved humankind, and only secondarily and consequentially the rest of creation.

Many problems with interpreting Romans 8:18–23 arise from failing to pay full attention to its context. It is, in fact, part of the theme of redemption and the Spirit's work which occupies Paul from chapters 5 to 8; the passage about the fall and suffering recalls 5:3–5, where suffering and hope are associated with the gift of the Spirit.

Now the fall resulted in death (Gn. 2:17; Rom. 6:23), that is, separation from God. This had two consequences: the relationship of 'love and cherish' between Adam and Eve became one of 'desire and domination' (*cf.* Gn. 3:16; 4:7); and 'tending' in Eden became 'toil' outside the garden (Gn. 3:18; *cf.* Lv. 26:3ff.; Pr. 24:30–34). Romans 8:20 states that the frustration currently experienced by creation is not innate in it, but was a consequence of 'the will of the one who subjected it'. Whether this was God or Adam does not matter; the point is that the frustration arises because of an extrinsic event, and can be dealt with by faith, as Paul pointed out in Romans 5:2. C. F. D. Moule paraphrases Romans 8:20: 'For creation was subjected to frustration, not by its own choice but because of Adam's sin which pulled down nature with it, since God had created Adam to be in close connection with nature' (1964: 11). The teaching of the whole of this central section of Romans is how Christ overcame death (on the cross), and how the consequences of this are dealt with, contrasting life in the flesh with life in the Spirit (Rom. 6:13). In Romans 8:19 Paul writes about the 'sons of God' who are to be revealed; in the same passage he defines 'sons of God' as 'those who are led by the Spirit of God' (8:14). The next verse (8:20) describes the vanity and frustration which result from a failure to respond to the Spirit. As Kidner says, 'leaderless, the choir of creation can only grind in discord' (1967: 73). The whole of the book of Ecclesiastes is a commentary on this verse.

The message of Romans 8:18–23 is thus one of hope – hope looking not only to the distant future but also to the time when the redeemed live out their reunion with God, and therefore their responsibility for nature. Paul's argument is that as long as humankind refuses (or is unable through sin) to play the role God created for it, the world of nature is dislocated and frustrated. Since humankind is God's vice-gerent on earth (which is part, at least, of the meaning of being 'in God's image'), we have inevitably failed in our stewardship from the moment we first disobeyed God and dislocated the relationship. Some Christians interpret any facts which they find morally difficult as the 'results of the fall' (such as 'nature red in tooth and claw', or the enormous number of human foetuses which spontaneously miscarry). We are almost completely ignorant about the moral situation before the fall. It is hazardous to argue from such apocalyptic passages as Isaiah 11:6–9 ('The wolf will live with the lamb') that particular ecological conditions were God's primary purpose.

Evolutionary ethics

There have been many attempts to show that ethics or moral standards have been formed by the evolutionary process itself, but such attempts have been uniformly unsuccessful. The difficulty is that natural selection means that individual survival and reproduction are rewarded, whereas altruism is punished. Darwin himself was aware of this. In *The Descent of Man*, he wrote:

> It is extremely doubtful, whether the offspring of the more sympathetic and benevolent parents, or of those who were the most faithful to their comrades, would be reared in greater numbers than the children of selfish and treacherous parents belonging to the same tribe. He who was ready to sacrifice his life, as many a savage has been, rather than betray his comrades, would often leave no offspring to inherit his noble nature. The bravest men, who were always willing to come to the front in war, and who freely risked their lives for others, would on average perish in larger numbers than other men. Therefore it hardly seems probable that the number of men gifted with such virtues, or the standard of their excellence, could be increased through natural selection, that is by the survival of the fittest (1871: 200).

The difficulty arises if we assume that all our characteristics are determined by the genes we inherit from our parents; Darwin's problem exists only if moral traits are inherited in a similar way to height or eye colour. Aldous Huxley must bear some of the blame for this assumption: in *Brave New World* (1932) he spelt out the Darwinian nightmare of matching (or condemning) people of different genomes to particular occupations. We now know that genetic determinism has been grossly over-exaggerated (p. 162).[9]

This is not the place to argue the correctness or otherwise of such (sociobiological) claims in humans. The arguments turn on the relative importance of the factors which control human behaviour. What is relevant here is to recognize that this debate can be conducted entirely on the scientific level, without Christians having to take up positions. The claim that sociobiology can replace (or control) ethics is wholly wrong (see *e.g.* P. Singer 1981); if we are an *imago Dei* (which we accept or reject by faith, not by proof in the scientific sense), then our behaviour is affected by it, even though we may have an entirely naturalistic explanation for the behaviour in terms of human history and natural selection. The debate is exactly the same as the one that took place in the eighteenth and nineteenth centuries when it was realized that our physiological processes are analogous to those of a machine, or in the 1930s and 40s when the brain was seen to have similarities to a computer. A generation ago, William Sargant discussed the signs associated with John Wesley's preaching and maintained that the conversions associated with the preaching were spurious, because the signs could be mimicked by physiological or pharmacological techniques (Sargant 1957). This claim was as mistaken as that of Singer: it assumes that any event can have only one description, and that knowing a causal mechanism excludes the intervention or control of God (p. 106). In all these cases the fallacy is to assume that we are 'nothing but' a machine, a computer, or a programmed genetic reaction; we *are* these, but we are *also* human beings made in the image of God.

There are three possible views about human origins: as 'nothing but' a highly evolved ape; as 'nothing but' a special creation of God made complete in every respect; or as an ape transformed by the conferring of God's image, with an evolutionary history and a unique relationship with the Creator. The important point to recognize is that only the last viewpoint (sometimes called 'theistic creationism') allows us to accept the validity of scientific investigation of our nature while still maintaining a personal potential of relationship to God.

The theistic creationism we have expounded involves specific interactions of God with creation on at least three occasions (the *bārā'* events), most importantly and contentiously with the origination of humankind. We could be accused of 'semi-deism', of smuggling Paley's divine watchmaker back into the creation process and compromising the principle that God upholds *all* things at *all* times (p. 40). Notwithstanding, most commentators accept that in the context of Genesis 1 there are differences between *bārā'* and *'āśâ*, with the former carrying the implication of creation out of nothing (although *'āśâ*, not *bārā'*, is used of the creation of Adam in Gn. 2:7) (Atkinson 1990: 21). The crux is how we should understand Romans 5. We follow John Stott in being persuaded that Paul's analogy between Adam and Christ depends for its validity on the equal historicity of both. But others interpret differently. For example, Bernard Ramm (1985) suggests that Adam was generic, not an individual. James Dunn goes further:

> It would not be true to say that Paul's theological point [in Rom. 5:12–21] depends on Adam being a 'historical individual' or on his disobedience being a historical event as such . . . So long as the story of Adam as the initiator of the sad tale of human failure was well known, which we may assume (the brevity of Paul's presentation presupposes such a knowledge), such a comparison was meaningful (Dunn 1988: 289).[10]

Such an exegesis, however, raises considerable problems about humanness and the fall. Nevertheless, the argument is wholly theological and does not affect the scientific understanding of palaeoanthropology (and the genetic continuity between *Homo sapiens* and other animals; Day 1998), or the behavioural differences between humans and non-humans, which are so large as to amount to a qualitative distinction (chapter 11).

Conclusions

In this chapter, we have discussed the interpretation of particular biblical passages in more detail than in any other chapter. Our reason for doing this is not that there is more in the Bible about evolution than on the other topics covered in this book, or that we are more confident in handling the biblical material on creation than that on other topics we deal with. It is simply that debates in Christian circles about evolution tend to be about the meaning and authority of the Bible. In one sense, we ought to put disputes about evolution and science behind us (after all, Paul commands Timothy 'to instruct certain people to give up teaching erroneous doctrines and devoting themselves to interminable myths and genealogies, which give rise to mere speculation and do not further God's plan for us' (1 Tim. 1:3–4, Revised English Bible; 'genealogies' are one way of looking at evolution), but in another, the activity of God in creation and his relationship to it form the crux of the scientific enterprise for the Christian. Consequently, we conclude this chapter by affirming some basic premises.

1. God created the world and everything in it.
2. Genesis (indeed the whole Bible) is little concerned with *how* God carried out his mighty works. The Bible does not describe *how* God created.
3. The scientific description of events is *complementary* and not contradictory to the Bible or to teleological explanation. This does not mean that either is superior or overruling.
4. There is no scriptural reason for disbelieving that God worked through

biologically understood mechanisms of evolution by natural selection to produce the world as we see it today.

5. Humans are qualitatively distinguished from the animals *only* in being made in God's image, which is relational and representational, not genetical or anatomical. Consequently, it is reasonable to suppose (and Gn. 2:7 suggests) that God 'placed his image' in an already existing animal.

6. The controversy over the Bible and evolution is intrinsically sterile. As often presented, it probably scares many young people away from seriously considering the person and claims of Christ. But the debate does entail three benefits. First, it teaches us to distinguish between the path of our understanding of Scripture which is derived from its true meaning and that which is uncertain in interpretation, based on uncertain or mistaken preconceptions. Secondly, it enables us to recognize our 'design limitations' with respect to everyday problems such as sex, Sabbath-keeping, family obligations, and especially obedience to God. These are sometimes called 'creation ordinances', and should be seen as such; they are not mere arbitrary regulations laid down by a distant despot. Thirdly, and most importantly, it forces us to recognize that God is active in 'normal', everyday events. In other words, God is relevant, active and powerful, completely distinct from the transcendent watchmaker of some creationists or the woolly immanent urge of the liberals.

J. B. S. Haldane spoke truly when he said, 'Science cannot give an answer to the question, "Why should I be good?"' Science cannot, but the Bible can.

Chapter 7

Evolution

Evolution is change (transformation) in the world since the original creation. It is firmly grounded in science. It developed as a science distinct from speculation only since it was realized (in the late eighteenth century) that the Earth is old; Darwin's chief contribution was to suggest a (testable) mechanism for evolutionary change. Many of the criticisms against evolution (which must be distinguished from unjustifiable evolutionism) are not valid. There is no necessary conflict between scientific understanding of evolution and the biblical revelation.

> Extinguished theologians lie about the cradle of every science as the
> strangled snakes beside that of Hercules; and history records that
> whereas science and orthodoxy have been fairly opposed, the latter
> have been forced to retire from the lists, bleeding and crushed if not
> annihilated; scotched if not slain (T. H. Huxley).

The confusions and passions of the debates between Christian faith and evolutionary science are a microcosm of the whole religion–science interface. On the one hand, Christians have been concerned to defend their understanding of the world as God's creation and upholding; on the other, an apparently never-ending procession of scientists have sought to show that the idea of a creator God is an 'unnecessary hypothesis'. Even more confusing have been disagreements among Christians, some espousing a very literal understanding of the Bible (that God created the world by fiat a few thousand years ago), and some (theistic evolutionists) claiming that there is no conflict between the Bible accounts properly interpreted and evolutionary science; while yet others reject scientific evolution on the grounds that methodological naturalism is an improper and insufficient ground for understanding God's creative work. Alvin Plantinga (1991) labels this third approach 'provisional atheism'. But as Ernan McMullin has pointed out forcefully, 'methodological naturalism does not restrict our study of nature; it just lays down what sort of study qualifies as scientific' (1993: 303).

McMullin reviews the categories of evidence for evolution (or the thesis of common ancestry [TCA], as Plantinga calls it) challenged by Plantinga (fossil record, molecular similarities, and the extent of variation), and draws from them support for consilience, that is, strengthening of the whole through independent lines of evidence:

> Evidence from three quite disparate domains supports a single coherent view of the sequence of branchings and extinctions that underlie TCA. If TCA is *false*, if in fact the different kinds of organisms do not share a common ancestry, this consilience goes unexplained. It is all very well to say, 'but God *could* have . . .' Such a hypothesis treats the consilience exhibited by TCA as a coincidence; it does not explain it. So it is not as though allowing the theistic alternative into the range of possible alternatives alters the balance of probability drastically, as Plantinga supposes. TCA is, of course, a *hypothesis*, as any reconstruction of the past must be. But it remains by far the best-supported response, for the theist as for others, to the fast-multiplying evidence available to us . . . When Augustine proposed a developmental cosmology long ago, there was little in the natural science of his day to support such a venture. Now that has changed. What was speculative and not quite coherent has been transformed, thanks to the labors of countless workers in a variety of different scientific fields. TCA allows the Christian to fill out the metaphysics of creation in a way (I am persuaded) Augustine and Aquinas would have welcomed. No longer need one suppose that God must have added plants here and animals there . . . (McMullin 1993: 319, 328).

The debate about naturalism has become fashionable in the 1990s. On the one hand it can be regarded as mere semantics; on the other it can be interpreted as an attempt to reopen the questions which stimulated the original edition of this book thirty years ago and their re-examination here. The second attitude sees the debate about evolution 'as a holy battle between two incompatible world views, one − theistic realism − that upholds truth, goodness and beauty, and another – evolutionary naturalism? – that undermines the same, leading to moral decay in every aspect of society, not to mention the loss of hope for an eternal afterlife' (Pennock 1996: 28, commenting on Johnson 1991; 1995). Meanwhile, the outsider looks on and is bemused, unconvinced of the reality of the debate and, more likely than not, confirmed in a belief about the irrelevance of Christianity in the scientific era.

What are the rights and wrongs of the situation? The traditional understanding of the Bible was that God made the world and all that is in it during a six-day period a few thousand years ago. Indeed, older Bibles often headed the marginal notes of the first chapter of Genesis with the date 4004 BC. This timing was calculated by Archbishop Ussher of Armagh (who rather appropriately was also Professor of Theological Controversies at Trinity College, Dublin) in a book (*Annals of the Ancient and New Testaments*) published in 1650, by the simple method of adding up the ages of all the people in the biblical genealogies from Adam to Christ.

He was not the first to attempt the sum: the Jewish calendar (still in use) dates the creation in 3761 BC, while the Venerable Bede placed it at 3952 BC. Rather more contentious was the idea of John Lightfoot, Vice-Chancellor of Cambridge University and a contemporary of Ussher's, who went further, and deducted that creation was completed at 9.00am on Sunday 23 October.

Ussher's arithmetic is perfectly sound. The difficulty is that it places Adam as living at a time when there was already considerable urban civilization in the Middle East, thus removing any credibility about his being the founder of the whole human race.

It is easy to condemn Ussher for naïveté, but we must not judge him by modern standards. His own answer to the question 'How do we know about creation?' was: 'Not only by the plain and manifold testimonies of Holy Scripture, but also by light of reason well directed.' His main quarrel was not with other timings of the human epic, but with Aristotle's unhistorical notion of eternity: 'What say you then to Aristotle, accounted of so many the Prince of Philosophers, who laboureth to prove that the world is eternal?' Ussher answered his own question by defending God's majesty against a mere unmoved mover of eternal matter, for Aristotle 'spoileth God of the glory of his Creation, but also assigneth him to no higher office than the moving of the spheres, whereunto he bindeth him more like a servant than a lord'.

The irony is that

> We castigate Ussher for making the creation so short – a mere six days, where we reckon billions for evolution. But Ussher feared that six days might seem too long in the opinion of his contemporaries, for why should God, who could do all in an instant, so spread out his work?

Why was he creating so long, seeing he could have perfected all the creatures at once and in a moment? Ussher gives a list of answers, but one [is] . . . 'To teach us the better to understand their workmanship; even as a man which will teach a child in the frame of a letter, will first teach him one line of the letter, and not the whole letter altogether' (Gould 1993: 193).

Notwithstanding, most people nowadays reject not only Ussher's date but the Genesis creation accounts in total; disbelieving one way of looking at one part of creation, such people tend also to throw away the creator who is revealed in the early chapters of the Bible and make humankind, created in his image, no different from any other animal.

Scientific background

In exactly the same way that the Ptolemaic cosmos of crystal spheres became incredible as astronomical knowledge grew in the fifteenth and sixteenth centuries, so the traditional understanding of the biblical story of creation became strained in the eighteenth and early nineteenth centuries. Up to the end of the seventeenth century the idea that 'natural' events could explain Earth history without deviating too far from a literal interpretation of Genesis was widespread. In his *Sacred Theory of the Earth* (1681), Thomas Burnet set out his version of the whole history of the Earth from creation to present. Particularly important to him was the flood, caused by a bursting of the outer crust of the Earth and an eruption of subterranean waters to the surface. Burnet believed that everything, including the final fires on the day of judgment, were due to natural causes that God had set in train at the creation. John Woodward, in his *Essay towards a Natural History of the Earth* (1695), was more conservative. The deluge was due to direct intervention by the Lord, but since then the world has again been more or less untroubled. All fossils are the product of the flood and prove its actual occurrence, thus confirming the account of the Bible. It was a very comforting interpretation. A third writer on the Earth's history, William Whiston, attempted to interpret the story of the Bible in terms of Newtonian physics. He speculated in his *New Theory of the Earth* (1717) that Noah's flood had resulted from a near approach of a comet.[1]

At the end of the eighteenth century, most geologists regarded the change between successive strata as the result of multiple catastrophes, the later forms being replacements specially made by the Creator. The hangover from Plato's ideas about the impossibility of biological change prevented the fossil record being interpreted as a sequence of continuing changes, as it would conventionally be seen today (Mayr 1982: 304).

The idea of evolution was in the air, as it were, during the second half of the eighteenth century. Maupertuis, Buffon and Diderot (Frenchmen), and Rodig, Herder, Goethe and Kant (Germans), have all been claimed as evolutionists, albeit without general support. They all postulated new origins or a simple 'unfolding' of imminent potential (rather than a change in an existing type). Nevertheless they are significant as being the immediate predecessors of Lamarck, who was the first to make a real break with the old, Plato-dominated worldview.

Lamarck

Jean-Baptiste Lamarck (1744–1829) was a protégé of Buffon, and from 1788 until his death worked in the French Natural History Museum. His key work was an evolutionary *Philosophie Zoologique* (1809), which he stated was needed to explain two phenomena: (1) the gradual increase in 'perfection' from the simplest animals to humankind, seen simply in terms of complexity, and (2) the amazing diversity of organisms. On these grounds, Lamarck claimed that one species may, over a longer period of time, become 'transformed' into a new species; such evolutionary change solved the problems of extinctions. In doing this, he introduced a time factor, which has been described as the Achilles' heel of natural theology: although a creator could design a perfect organism in an unchanging world, this would be impossible if the environment was changing, sometimes drastically (Ospovat 1981; Mayr 1982: 349). Adaptation to changes of climate, of the physical structure of the Earth's surface, and of predators and competitors could be maintained only if organisms could adjust to the circumstances; in other words, if they could evolve.

Lamarck was a uniformitarian: he believed that the same factors had been operating on the Earth from its beginning to the present time. This meant that he accepted that the Earth was very old, with organisms changing constantly, albeit very slowly. Furthermore, his studies in the Museum led him to believe that lineages did not only change in their own right, but that they also branched to produce separately evolving lines. The weakness lay in his suggestions about the mechanism whereby such transformations could come about. He believed there was a 'natural law' which produced a trend towards greater complexity 'from powers conferred by the "supreme author of all things" . . . Could not this infinite power create an order of things which gave existence successively to all that we can see as well as to all that exists but that we do not see?' He also believed that individuals had a capacity to react to that environment, so as to remain in perfect harmony with it, although he did not state how newly acquired characteristics were supposed to be inherited. He simply asserted that 'New needs which establish a necessity for some part really bring about the existence of that part as a result of efforts'. This was very vague.

Lamarck prepared the way for Darwin in pointing to evidence that evolutionary change must have occurred, but the two men's approaches and contributions were utterly different. Fifty years after Lamarck, Darwin wrote that *Philosophie Zoologique* was 'veritable rubbish . . . I got not a fact or idea from it', although on another occasion he acknowledged: 'The conclusions I am led to are not widely different from his [Lamarck's]; though the means of change are wholly so.' The crucial difference between Lamarck and Darwin was that Lamarck believed that the environment and its changes had priority, since these produced needs and activity for organisms, which in turn caused adaptational variation; in contrast, Darwin began with the existence of variation in organisms, and the ordering action of the environment ('natural selection') followed.

From Lamarck to Darwin

Before we reach Darwin, however, much more happened. Particularly significant was the attempted reconciliation of geology with Genesis. Superficially, science and the Bible agreed in that human remains (the last act of creation) were apparently absent from the fossil record, and in that there seemed to be evidence of a great deluge covering the whole Earth. As early as 1681, Thomas Burnet calculated in *The Sacred Theory of the Earth* that eight times the volume of water in the present oceans would have been needed to cover all the land areas on Earth. Burnet believed that all this water came from undersea caverns. Notwithstanding, by the eighteenth century the repeated changes in fossil faunas led to proposals for a whole series of floods.

In 1830, Charles Lyell published the first volume of his *Principles of Geology*, a book which had a great influence on Darwin. For Lyell, all geological processes were the results of secondary causes, that is, they did not require interventions. He was a 'uniformitarian', in contrast to the prevailing 'catastrophists', who assumed that God had repeatedly created new species after the recurrent catastrophes revealed in the fossil record. For most of his life Lyell was a firm opponent of evolution, although he was ultimately persuaded by Darwin.

Then in 1844 a book was published in London which blasted the genteel debates of the time. It sold more than twice as many copies in the ten years after its publication than either Lyell's *Principles* or Darwin's *Origin. The Vestiges of the Natural History of Creation* was so heretical that the author took every precaution to remain anonymous; speculations about his identity ranged from Lyell or Darwin to Prince Albert, Queen Victoria's husband. Adam Sedgwick, Professor of Geology at Cambridge and Darwin's teacher, exploded in April 1845 with a review that stretched to eighty-five pages. He wrote to Lyell: 'If the book is true, the labours of sober induction are in vain; religion is a lie, human law is a mass of folly, and a base injustice; morality is moonshine; our labours for the black people of Africa were works of madmen; and man and woman are only better beasts' (cited by Gillispie 1951: 165).

The author of the *Vestiges* was Robert Chambers, although this became known only after his death in 1871. He was a popular essayist and the editor of *Chambers' Encyclopaedia*. He took an avowedly Christian stance, believing that when there is a choice between special creation and the operation of general laws instituted by the creator, 'the latter is greatly preferable as it implies a far grander view of the divine power and dignity than the other'. He had no doubt that the available fossil evidence showed that the fauna of the world had evolved through geological time, and, since there is nothing in inorganic nature 'which may not be accounted for by the agency of the ordinary forces of nature', why not consider 'the possibility of plants and animals having likewise been produced in a natural way?'

What Chambers did was to apply uniformitarianism to organic nature. The hierarchy of animals made no sense to him unless one adopted evolution. Like Darwin, he constantly emphasized how many phenomena, for instance vestigial organs, could be explained as the product of evolution, while they had no easy place in terms of special creation.

The trouble was that Chambers made many mistakes. He believed in spontaneous generation and backed up his belief by all sorts of folklore. He advocated no real mechanism whereby change might occur. He was savaged by his critics. But he converted some influential people, including A. R. Wallace, the philosophers Herbert Spencer and Arthur Schopenhauer, and the poet Ralph Waldo Emerson. More importantly, he accustomed people to the idea of evolution; critics of the *Vestiges* supplied the standard objections to evolution, which Darwin took care to answer in the *Origin*.

Darwin

And so we come to Charles Darwin (1809–82), younger son of a country doctor and one-time intending ordinand. It was Darwin's *The Origin of Species by means of natural selection, or the preservation of favoured races in the struggle for life* (first published 1859, sixth and final edition 1872) that led to the acceptance of evolutionary ideas by both the scientific and the general world. The reason for the immediate success of the *Origin* was Darwin's explanations for the distribution of animals and plants, and his convincing interpretation of the significance of vestigial organs. Other lines of evidence, from fossils, anatomical likenesses, and so on, were well known to Darwin's contemporaries, but were not compelling to them because no acceptable mechanism of evolution was available. There is no doubt that it was Darwin's easily understood *mechanism* of evolution which was his most important contribution. A good example of the need for a mechanism before a scientific idea is generally accepted also occurred with continental drift, which was put forward in detail by Wegener in 1915, but not commonly accepted until the nature of tectonic plates was described by geophysicists in the 1960s.

Darwin's doubts about the immutability of species came from his study of geographical variation in both fossils and living forms during his time as naturalist on board the survey ship *Beagle* (1831–6). In 1838 he read 'for amusement' Malthus's *Essay on the Principle of Population*, and, 'being well prepared to appreciate the struggle for existence . . . it at once struck me that under these circumstances favourable variations would tend to be preserved, and unfavourable ones to be destroyed. The result of this would be the formation of new species. Here then I had at last got a theory by which to work.'

Darwin's original intention was to write a definitive book on evolution. In the spring of 1858, however, he was sent an essay, *On the tendency of varieties to depart indefinitely from the original type*, written by Alfred Russel Wallace while recovering from fever in the Moluccas. Darwin felt that this should be published at once, but on the urging of his friends Charles Lyell (whose *Principles* had first alerted Darwin to the reality of long-continued gradual geological change) and J. D. Hooker (son of the effective founder of the Royal Botanic Gardens at Kew, and instigator of the study of plant geography), he allowed a revised version of an essay he had drafted as early as 1844 to be forwarded with it to the Linnean Society.

The papers of both Darwin and Wallace clearly contain the three facts and two conclusions which are commonly taken as the simple summary of Darwinian evolution: the potential of all species to increase greatly in numbers, coupled with an approximate constancy of numbers, implies that there is a struggle for existence; and when variation is added to this, it is clear that natural selection must operate.

It was the ease with which these propositions could be understood that helped evolution to be generally accepted. In fact, Darwin devoted more than half of the *Origin* to different lines of evidence that evolution had occurred: he had two chapters on the fossil record, two on geographical distributions, and one each on morphological likenesses (including comparative embryology, the interpretation of vestigial organs, and the meaning of classification), behaviour, and domestication. He devoted later books specifically to the origin of humankind and sexual selection, domestication, and adaptations in plants for pollination, insect-eating, and climbing. All these were parts of the book which Darwin had originally intended to write before being forced into print by Wallace.

In chapters 6 and 7 of the *Origin* Darwin dealt with 'difficulties' and 'miscellaneous objections' to his theory. His main points concerned the nature of species and questions about the efficacy of selection. In a later chapter, he discussed the imperfections of the fossil record. Darwin knew that the maintenance of variation was a key weakness in his theory. The causes of variation are repeatedly referred to in the book, and in later editions of the *Origin* he tended to accept that some Lamarckian explanation might be necessary (*i.e.* that the heredity of an individual might be affected by an environmental modification of its phenotype. No claim of such 'Lamarckian' inheritance by Kammerer, Lysenko, Steele, or many others over the years, has, however, ever been substantiated). The problem was not resolved until the physical basis of heredity was understood following the embryological conclusions of Weissman (1883) and the rediscovery of Mendel's work in 1900 (Mayr 1982: 629ff.; 1991).

The issues that Darwin faced in the *Origin* are still raised today; it is one of the irritating and frustrating elements of evolutionary debate that old questions are repeatedly posed, but the old answers not given.[2]

The effectiveness of natural selection

The most persistent criticism of Darwinism has always been that natural selection is merely a negative instrument removing inefficiency, but incapable of producing novelty or the seemingly perfect adaptation of such features as the eye of a mammal or bird, or the pattern of a butterfly's wing. This is an area where evolutionary biologists have shown that the old objections are weak (see *e.g.* Dawkins 1996), but their conclusions were to some extent anticipated by Darwin himself. It is worth dealing with the criticism because it so often resurfaces. We can deal with it under three headings:

Cases of special difficulty

Darwin recognized three situations here:

1. An organ (such as the wing of a bat) may be so specialized for its functions as to bear little resemblance to the prototype (the forelimb of a land-living insectivore) from which it must be presumed to have arisen. The difficulty is to envisage a series of organisms with organs of intermediate grades connecting these widely separated extremes.

2. An organ of extreme perfection (such as the eye of higher vertebrates) may show such perfect and detailed adaptation that, by comparison with the obstacles which the design of such an apparatus would present to human ingenuity, the mind is staggered by the effort of conceiving it as the product of so undirected a process as trial and error.

3. Some organs of seemingly trifling importance (such as the 'fly-whisk' tail of the giraffe) are yet so clearly adapted to the function they perform that they cannot be regarded as accidental. In these cases it may be asked how such a trifling function can ever have been a matter of life and death to the

organism, and so have determined its survival in the struggle for existence.

The first of these classes of objection applies to all evolution, while the second and third are difficulties more of imagination than of reason. R. A. Fisher has commented that 'the cogency and wealth of illustration with which Darwin was able to deal with these cases was, perhaps, the largest factor in persuading biologists of the truth of his views' (Fisher 1954: 89). Here we can note only two points. First, *function as well as structure evolves*. For example, there are organisms which have no image-forming eyes, but possess light-sensitive cells. Any inherited variants which allow detection of the direction of a light source, its size, movement, and so on, could be of potential advantage, and subject to natural selection. The eye as we know it would be built up by the accumulation of many small steps, each of them adaptive. The result is efficiency in a particular environment, not perfection: one could envisage the human eye being 'improved' by functioning better in poor light or under water, or failing to deteriorate with age; these attributes have never been 'necessary' for human survival.

Secondly, *the usefulness (or 'adaptive value') of a character can be tested by experiment*. Victorian biologists wasted a vast amount of effort speculating about the function or value of particular organs; there was a similar tendency in the 1980s to pontificate about the significance of particular behaviour patterns. Apparently trivial traits may be shown to be highly important; it has only recently been shown that flies may seriously disturb tropical herbivores, and an efficient 'fly-whisk' may add notably to fitness. Conversely, other traits (such as the horns of the Irish elk) may be incidental results of selection for other traits associated with growth.

The persistence of these criticisms has come from ignorance, often resulting from lack of research, rather than from defects in the underlying ideas.

The origin of novelty

How can natural selection, which functions to 'filter out' deleterious variants, lead to completely new developments? Is not natural selection limited to modifying existing adaptations? The answer is no, as consideration of three facts reveals. First, all traits are subject to variation. Secondly, in evolution, novelty is introduced by opportunity, usually arising through a change in the environment: animals and plants invaded land because of an available habitat, not because it seemed a good idea. Some characters will be pre-adapted to the new environment. Thirdly, even a very small selective advantage can lead to genetical change.

In the *Origin*, Darwin quoted Agassiz's work on echinoderms, showing how modification of spines may lead to the development of an apparently new and important trait, tube feet. Many similar examples are known. One of the major biological discoveries of the 1960s was the enormous amount of variation present in virtually every population, which means that a species can respond rapidly and precisely to new environmental stresses: bacteria can digest oil, aphids detoxify artificial poisons, and plants grow when introduced to Antarctica. These are described in all modern biological texts (*e.g.* Roughgarden 1979; Maynard Smith 1989; Berry, Crawford & Hewitt 1992; Skelton 1993). There is no reason for believing that any reasonable novelties cannot occur in evolutionary time. Indeed, even apparent disadvantages (such as sterile castes of insects) can evolve in appropriate conditions; natural selection is a mechanism for producing *a priori* improbable contingencies (Fisher 1954: 91). (This fact incidentally answers claims that there has not been enough time for evolution to have taken place since the Earth became habitable. The commonly used analogy that a monkey randomly typing will produce the works of Shakespeare, but only if he had astronomical time, is irrelevant, since selection can rapidly and ruthlessly change the frequency of apparently random variation.)

The strength of natural selection

The power of natural selection to produce adaptation could be illustrated only by anecdote until quantitative methods of estimating fitness differences, rate and conditions of gene frequency change, and similar parameters, became available. These were developed by R. A. Fisher, J. B. S. Haldane and Sewall Wright during the 1920s, and demonstrated the ability of selection differentials of less than 1% to bring about evolutionary change. The force of their theoretical arguments has been greatly strengthened with the discovery that selection pressures in nature commonly reach 10% or more.

The answers given by Darwin to his critics in successive editions of the *Origin* have subsequently been proved right by subsequent research. The rediscovery of Mendel's work in 1900 and the realization that genetic mutation formed the raw material for evolution ought to have solved Darwin's problem about the origin of variation; in fact it produced a rift between biometricians and palaeontologists on the one hand, and geneticists on the other. During the 1920s, the long-continued, gradual, evolutionary progressions revealed by study of fossils evoked theories of intrinsic evolutionary urges, linked with pantheistic notions of *élan vital* and 'emergent evolution'. These speculations were soon to be recognized as wrong, but they became entangled with philosophical examinations of progression and advance. For example, Alfred North Whitehead (1926) suggested that the world is best seen not as a collection of distinct objects, but as a complex of ongoing processes in which God is involved in every event, 'luring' the process in the direction he desires for it; his ideas became dignified and were developed as 'process theology' (see p. 101).

A moral objection to natural selection is that it involves waste and suffering which are not compatible with the care and design of a loving God. There is no dispute about this, as seen from a human perspective. But this is the wrong way to view the process. Physical death should not be equated with evil; it was part of God's plan from the beginning, when he made plants to be the food of animals. Plant death is as much biological death as animal death. We extrapolate 'suffering' into 'nature red in tooth and claw', whereas God's concern is separation and alienation from him. At one level we cannot understand 'natural wastage'; at another level, we can only acknowledge that God's ways are not ours, and that we must allow him to decree the best way to produce the end result he desires, whether or not it is wasteful in our eyes. We must beware of extrapolating our sentiments to God. It is certainly wrong for us to abuse animals and plants, by causing unnecessary suffering or killing them wantonly, because we are moral beings answerable to God for the way we treat his creation. It is wrong, however, to apply moral categories to the ways other living creatures treat each other. Our unease with the way that God has chosen to work is no scientific ground for criticizing evolution.

The essential correctness of Darwin's original ideas was scientifically recognized in the 1930s, beginning with R. A. Fisher's demonstration that there is no antithesis between genetic adjustment and gradual change. This led to the neo-Darwinian synthesis, which showed the irrelevance of internal urges (as described above) and which amounted to evolutionary pantheism. However, the errors of the 1920s still persist because they were incorporated into standard histories of biology written at the time (by Nordenskiöld 1928; Radl 1930; and C. Singer 1931), and are perpetuated by writers quoting these source texts. There are proper scientific debates about the evolutionary process, but they are about details of the relative importance of different mechanisms and rate (for example, the controversies about 'punctuated equilibrium');[3] they are not about whether or not evolution has occurred at all.

Evolution and Genesis

The Genesis accounts of creation are about the Creator first and foremost; they are only secondarily about his actions. It is God who speaks; it is God who sees that his work is good; and it is God who made us in his image. The author of Revelation sums up the position from the other end, as it were, of the Bible:

> You are worthy, our Lord and God,
>   to receive glory and honour and power,

for you created all things,
and by your will they were created
and have their being (Rev. 4:11).

The Westminster Confession makes the same point: 'God created the world for the manifestation of the glory of his eternal power, wisdom and goodness.' Calvin described creation as the theatre of God's glory (cited in *NBD* 245). As we have seen, creation was not necessary. God did not 'have' to create the world. 'The God who made the world and everything in it . . . does not live in temples built by hands. And he is not served by human hands, as if he needed anything, because he himself gives all men life and breath and everything else' (Acts 17:24–25). Both the world and humankind originate from the free and sovereign will of God. From the point of view of creation, the Creator is in every respect independent of his creatures. This completely distinguishes the Creator of the Bible from any pantheistic ideas which say that the living world represents expressions of God; 'for from him and through him and to him are all things' (Rom. 11:36).

The first three chapters of Genesis set the scene for the rest of the Bible, teaching about the nature of God and the impact of evil on the creation. We are presented with the outline of a great range of divine truths – about relationships with God and with each other, and about responsible behaviour. In theological language, Genesis 1–3 is a 'theodicy', stating and justifying God's goodness in an evil world. The problem we have to face is how to translate the language of these chapters for the modern day. We have to be extremely careful about our *interpretation* of these chapters; our prime concern must be what God is trying to say to us. Four general points are relevant.

First, the Genesis 1 and 2 accounts are different from each other, but complementary rather than contradictory; Von Rad sees them as 'in many respects open to each other', and argues that exegesis should be carried out on both of them together (1961: 41). The second story is in many respects a recapitulation of and sequel to the first account, and provides an introduction to chapter 3. In both accounts there is one supreme God, by whose act and word order was established out of chaos, and upon whom humankind is dependent for its existence and place in the order of created beings; both emphasize obedience to God and fellowship with him; and both introduce ideas integral to the whole of the Old Testament. Kidner underlines this: in chapters 2 and 3,

. . . man is the pivot of the story [while] in chapter 1 he was the climax. Everything is told in terms of him; even the primaeval waste is shown awaiting him (2:5b) and the narrative works outwards from man himself to man's environment (garden, trees, river, beasts and birds) in logical as against chronological order, to reveal the world as we are meant to see it: a place expressly designed for our delight and discipline. It is misleading to call this a second creation account, for it hastens to localize the scene, passing straight from the world at large to 'a garden in the east'; all that follows is played out on this narrow stage (Kidner 1967: 58).

Secondly, the Bible's Creator is distinct from those portrayed in other middle-eastern mythologies. He is presented as a living God, unmistakably personal. The verbs of Genesis 1 express an independent energy of mind, will and judgment which excludes all question of God's being 'it' rather than 'thou'. Marduk, Asshur, Baal, and others were members of pantheons. The biblical Creator is the only God, sovereign over all that is. The world is separate from him, not an emanation or expression. We have already seen the importance of this in the development of science (pp. 21, 63); a Jesuit writer, Robert Faricy, shows how it leads to a 'de-divinization' of nature:

> Genesis underlines that creatures are in no way divine . . . They are merely creatures, not divine, infinitely distant from their ineffably transcendent Creator, and completely subject to him as their Lord . . . Behind the earth's fertility, and causing it, we do not find the sun or the moon, or as in many myths, the tree of life. We find only the creative word of God. Nature is radically distinct from God (Faricy 1982: 3).

His ways are perfect. The series of cataclysms in Genesis declares that heaven can make no truce with sin, whether it is the Godward sin of unbelief and presumption (as in Eden and Babel), or the humanward wrongs of violence, lust and treachery. There is an Egyptian account dated *c.* 2350 BC which describes the god Atum who brought forth gods on a primeval hill above the waters of Chaos. Atum, 'who came into being by himself', next brought the world into order and, out of the dark deep, assigned places and functions to the other deities. Another myth describes humankind as being created from the tears of the sungod Ra, all people being created equal in opportunity to enjoy the basic necessities of life (*NBD* 248). The best known of the creation myths is a Babylonian adaptation of a Sumerian cosmogony known as *Enuma Elish*. This begins with two gods, Tiamat and Apsu, but then other gods were born, and Apsu tried to do away with them because of their noise. One of them, Ea, managed to kill Apsu; then Tiamat, bent on revenge, was killed by Ea's son Marduk. Marduk used the two halves of Tiamat to create the firmament of heaven and earth. He then set in order the stars, Sun, and Moon, and lastly, with the help of Ea, created humankind from clay mingled with the blood of Kingu, the rebel god who had led Tiamat's forces (*NBD* 247).

Throughout the ancient Near East there was a conception of a primary watery emptiness and darkness, and man made for the service of the gods. But the Old Testament account stands distinct with its clarity and monotheism; in it are no struggles between deities to exalt any particular city or race (*NBD* 248). The contrasts between Genesis 1 and all the known extrabiblical cosmogonies are more striking than the resemblances.

Thirdly, the Bible's creation narratives must not be read as *scientific* accounts in the modern sense (p. 28). This is not to impute factual inaccuracy, but to insist upon the purpose of the passages. Just as Galileo was right to insist that the Bible tells us how to go to heaven, not how the heavens go, so we must beware of reading biology into the creation narratives when their primary aïm is theological. From an avowed 'creationist' standpoint, Cameron (1983: 49, 61) has argued that since Genesis 1 – 3 is a theodicy, all attempts to harmonize scriptural and secular language must fail because they are irrelevant in the context.

The most persistent misapprehension about God and creation, however, is that knowledge of causal mechanisms automatically excludes any possibility that God is acting in a particular situation. This brings us back to the core of the book, and the need to appreciate the complementarity of scientific (or causal) and formal explanations, which may involve divine activity. The latter can be approached only by faith. ('By *faith* we understand that the universe was formed by God's command, so that the visible came forth from the invisible'; Heb. 11:3, REB). It is this complementarity that Richard Dawkins does not accept in his book *The Blind Watchmaker* (1986) and other writings. Although he is correct to insist that a formal explanation of a cause is not necessary, he is logically incorrect to argue that it can therefore be excluded.[4]

What is the connection between God and creation? The best way of regarding divine creation and biological evolution is to understand them as complementary explanations. The God of the Bible is primarily a creative upholder (Col. 1:17), and only secondarily a divine watchmaker. There is no conflict or rivalry in distinguishing between *why* God created and the methods or mechanisms used, which are the business of science to probe. Francis Schaeffer described how his approach to Genesis 1 changed as he reflected on God's relationship to his creation:

> As a younger Christian, I never thought it right to use the word *creation* for an artist's work. I reserved it for God's initial work alone. But I have come to realise that this was a mistake, because while there is indeed a difference there is a very important parallel. The artist conceives in his thought-world and then he brings forth into the external world . . .

And it is exactly the same with God. God who existed before had a plan, and he created and caused these things to become objective (Schaeffer 1973: 27).

In other words, we must distinguish between *why* God created (which is described in the Bible) and the mechanism he used, which is the role of science to investigate. The fact that God created all matter and life, and did not merely shape it, is important. It is implicit in some of the ideas of the previous two sections. Prior to the action of God 'in the beginning' (Gn. 1:1), there was no existence other than God's. This means that matter is not eternal, and that there is no other power in existence in the Universe outside God's control. It indicates also, as we have already seen, that God is distinct from his creation, which is not merely an external manifestation of an Absolute. There are several strands of evidence for this doctrine of creation from nothing (*ex nihilo*) in the Bible.

First, there is no mention of any pre-existing material out of which the world was made. (Although it is possible to understand Gn. 1:1–2 as indicating that God ordered previously formed matter, this contradicts statements elsewhere – *e.g.* Ps. 33:6–9; Is. 48:12–13; Col. 1:15–17 – that the whole material world was brought into being by God.) The Scriptures never represent the world as an emanation from God by a necessity of his nature.

Secondly, the descriptions of primary creation rule out any idea of mere ordering. In Psalm 33:6 we read, 'By the word of the LORD were the heavens made.'

Thirdly, the same doctrine is involved in the absolute dependence of all things on God and in his absolute sovereignty over them. Thus Ezra addressed God:

You alone are the LORD. You made the heavens, even the highest heavens, and all their starry host, the earth and all that is on it, the seas and all that is in them. You give life to everything (Ne. 9:6; see also Col. 1:16–17; Rev. 4:11).

Everything other than God is said to owe its existence to his will. The author of Hebrews, as we have seen, begins his illustration of the nature and power of faith by referring to creation as the great fundamental truth of all religion (Heb. 11:3). Creation as a divine act is a fact which we know only by revelation.

Finally, the doctrine of creation derives from the infinite perfection of God. There can be only one infinite being. If anything exists independently of his will, God is thereby limited. The God of the Bible existed outside of and before the world, independent of it, its creator, preserver and governor.

The doctrine of creation is a necessary consequence of theism (Ward 1996: 79 *et passim*). Hence the doctrine is presented on the first page of the Bible as the foundation of all subsequent revelations about the nature of God and his relation to the world, and from the beginning one day in seven is appointed to be a perpetual commemoration of the fact that God created the heaven and the earth.

Chapter 8

Biblical portraits of human nature

It is not only biologists and psychologists who offer us portraits of human nature. The world's religions present a spectrum of views of human nature. The Bible is full of teaching about human nature, and biblical scholars are helping us to understand more of the variety of biblical portraits of ourselves.

In this chapter, we focus on the Hebrew-Christian pictures of human nature. How do the Scriptures encourage us to think about ourselves? Do they present, for example, a dualistic view — soul and body held together in some undefined way? Or do they compartmentalize us into distinct substances: soul, mind, body, heart, spirit, and so on? And if so, how are these interrelated, and how do they interact? And what happens at death? How do we adhere to the biblical doctrine of resurrection and avoid slipping unthinkingly into pagan views of immortality? And what about the spiritual dimension of humankind?

At this point, it is pertinent to return from specific scientific issues to the Bible and to ask what it says about the nature of human beings, their life and death. Scientists and philosophers tend to develop their own ideas about these. For example, both Karl Popper and John Eccles (1966) argue that human nature is 'dualistic', made up of two distinct substances. Others are unashamedly reductionist. Francis Crick asserts that 'you, your joys and sorrows . . . are no more than a vast assembly of nerve cells and their associated molecules . . . the idea that man has a disembodied soul is as unnecessary as the old idea that there was a Life Force' (1994). Richard Dawkins (1976) tells us that we are nothing but survival-machines for genes.

The background

The formation, emergence and changes in Christian views of human nature over the centuries are complex, but everyone agrees that they have been significantly influenced by ancient philosophy, and this in turn has affected our understanding and interpretation of the Bible. For example, the Greeks typically began from the assumption that the human being is a soul temporarily imprisoned in a body (that is, without physical form). On this view the soul conferred consciousness and was immaterial and eternal. Plato claimed that it was made up of three 'parts': reason; the spirited element, which initiates action; and an element producing drives and appetites.

The Neoplatonists elaborated Plato's ideas and incorporated them into religious systems. Augustine (354–430) believed that a human being is a rational soul *using* a mortal and material body rather than simply being imprisoned in the body. He retained the tripartite idea but changed the nature of the parts. Probably the commonest idea of the will is still the Augustinian notion, and it, rather than reason, has come to be seen as the dominant aspect of the soul.

Thomas Aquinas (1225–74) added the ideas of Aristotle (384–322 BC) to the prevailing views. For Aristotle, the soul was not so much an entity as a life principle; while plants and animals have nutritive and sensitive souls, there is also an aspect of the soul which provides the powers or attributes characterizing the human being. Since the soul is a principle of body functioning, it dies with the body, although Aristotle speculated that some aspect related to rationality might survive death. Aquinas believed that this was helpful in understanding Christian teaching on the resurrection of the body, although that doctrine required that there must be an immortal soul to which the body would be restored at the general resurrection. He thus differed from Aristotle about the mortality of human souls.

Effectively, both Greek and medieval theology recognized that human beings have some specific human capabilities (such as the ability to pursue mathematics and philosophy), while others they share with animals (such as sensation). Since Aquinas and his followers believed that it was difficult to attribute the former to the body, they assumed that there must be another component of the person. Living persons can do all these things and dead ones cannot, so the soul was identified with the life principle.

As our knowledge of anatomy and physiology has grown, so the problems of accounting for the nature of the human person have become increasingly acute. How can a non-material entity, the soul, interact with a material body? We have implicitly absorbed the distinction made by René Descartes (1596–

1650) of two basic kinds of reality: extended substance (material things) and thinking substance (God, angels, human minds). Note the shift from 'souls' to 'minds': Latin has two words, *anima* and *animus*, each translated 'soul', but *animus* is also translated 'mind', while *anima* means the principle of life. English-language philosophy has tended to use the term 'mind' rather than 'soul' in most cases. It is against this background that the problem of mind–body interaction has become more difficult as Aristotelian conceptions of matter have been succeeded by revised versions of atomism.

Partnership in understanding

The questions faced by twentieth-century scientists were completely unknown to the biblical authors. The Bible does, however, have many things to say about human nature. Classic passages such as Psalm 8 pose the question, 'What is man that you are mindful of him?' (verse 4), emphasizing the *theocentric* context of scriptural statements, which are primarily to do with humankind *in relation to God*, creator, sustainer and redeemer. A major concern for us must be the meaning of the Bible passages to the original readers. This means focusing on such questions as whether an immortal soul is implanted in each of us at birth; whether this soul will be released from its temporary physical embodiment at our physical death; whether there is within each of us some non-physical part which remains immune from the 'changes and chances of this mortal life', whether we have a personal 'soul' which, like a pilot flying an aircraft, steers our body through life's varied calm and turbulent conditions; and whether, if there is such a thing as a 'soul', where and how it exerts its controlling influence over our brains, and thence our minds and bodies.

We may more easily avoid gross errors if we remember that 'it is never wise to bring to a passage of Scripture our own ready-made agenda, insisting that it answers *our* questions and addresses *our* concerns. For that is to dictate to Scripture instead of listening to it. We have to lay aside our presuppositions, so that we can consciously think ourselves back into the historical and cultural settings of the text.' These are John Stott's words (1994: 189), and implementing them will allow us better to 'let the author say what he does say and not force him to say what we want him to say'. It is up to us to search the Scriptures.

Human nature: a partnership of complementary accounts

Insights into human nature – both profound and trivial – are recorded in literature, portrayed in art, proclaimed by religions, and in the majority of cases long pre-date scientific discoveries. For good exegetical reasons, few people today regard statements in the Bible about the creation of the Universe as competing with what geologists, geophysicists and astronomers

138 Science, life and Christian belief

tell us about the origins and ages of the Earth, the Sun, the planets and galaxies. Nor do we seek to incorporate either into the other; that way lies confusion and an abuse of Scripture. We labour this point in the context of understanding human nature because too often we try to force the Bible categories of mind, soul and so forth into modern scientific concepts.

Bluntly, we are in danger of doing what our predecessors did to Galileo when convicting him of heresy on the basis of their constrained interpretation of Holy Writ. After all, was it not clear in Galileo's day that the Earth was fixed, immovable and non-rotating (Jos. 10:12; 1 Ch. 16:13; Ps. 93:1; 96:10; 104:5)? Is it not as obvious to us that humans possess immortal souls? That we are made up of two parts? Or three parts? That we are distinguished from the animals by possessing a soul?

The problem we face in relating biblical statements about human nature with those made by scientists is compounded when compared with those in the physical and earth sciences, because the words used are largely the same, particularly when we come to theories of personality. Indeed, the pronouncements of some psychotherapists sound remarkably like the statements that we find about human motivation and human nature in the Bible. For that reason alone it is not always immediately evident that scriptural statements differ radically from what we read in psychologists' writings about human nature.

Granted that the concerns of biblical writers are self-evidently different from those of psychologists, it nevertheless remains the case that both share a concern about the components of human personality. The problem is that theological and psychological models appear, at times, to be competing: is the individual a package composed of several parts, referred to in the Bible as soul, mind, heart, spirit and body; or is he or she a single mind/brain unity as described in psychology and neuroscience? How are the parts making up the whole interrelated? Such questions have fascinated humankind down the centuries.

The biblical portrait of human nature
A contemporary issue

The need to set out the main features of human nature derivable from a study of the *whole* of Scripture is particularly urgent because some leading scientists use their definition of the Christian view of humankind as a straw man, and then solemnly dispose of it as untenable. For example, Francis Crick opens his book *The Astonishing Hypothesis* (1994) with the Roman Catholic catechism's question, 'What is the soul?' The answer is: 'A soul is a living being without a body having reason and free will.' Crick sees this as a Christian assertion that 'people have souls, in the literal and not merely the metaphorical sense'. He then goes on to demolish any such view.

In fairness to Crick, his failure to ask 'What is the biblical view of human nature in general, and of the soul in particular?' is shared by many Christian writers. For example, both Martin Luther and William Tyndale list as the last of the 'five cardinal errors of the papal church' the doctrine of the natural immortality of the soul. As we have seen, this doctrine is an assumption based on Aquinas following Aristotle; we shall go on to show that it is not an essential or necessary element of Bible interpretation.

What are the main lines of scriptural teaching on the make-up of human nature? How are the words translated as 'soul', 'spirit', 'body' and 'heart' used in Scripture, and what composite pictures do they give us about our nature? The need to go to the Bible in arriving at a view of human nature is important because failure to do this can spark conflicts between science and faith, and, more insidiously, can lead to Christians erroneously defending 'accepted views' which are not biblical ones.

B. O. Banwell, in his article on 'Body' in the *New Bible Dictionary*, tells us that 'contrary to Greek philosophy and much modern thought, the emphasis in Hebrew is not on the body as distinct from the soul or the spirit' (*NBD* 145). Writing on 'Heart', he says it 'was essentially the whole man, with all his attributes, physical, intellectual and psychological, of which the Hebrew thought and spoke, and the heart was conceived of as the governing centre for all of these . . . Character, personality, will, mind are modern terms which all reflect something of the meaning of "heart" in its biblical usage . . . "Mind" is perhaps the closest modern term to the biblical usage of "heart", and many passages . . . are so translated' (*NBD* 465). Of the 'soul' we read, 'Usually the *nepeš* [soul] is regarded as departing at death (*e.g.* Gn. 35:18), but the word is never used for the spirit of the dead' (*NBD* 1135). And J. E. Colwell's *New Dictionary of Theology* article on 'Anthropology' says that

> Gn. 2:7 refers to God forming Adam 'from the dust of the ground' and breathing 'into his nostrils the breath of life', so that man becomes a 'living being'. The word being [*nepeš*] . . . should . . . be understood in its own context within the OT as indicative of men and women as living beings or persons in relationship to God and other people . . . According to Gn. 2, any conception of the soul as a separate (and separable) part or division of our being would seem to be invalid. Similarly, the popular debate concerning whether human nature is a bipartite or tripartite being has the appearance of a rather ill-founded and unhelpful irrelevancy. The human person is a 'soul' by virtue of being 'body' made alive by the 'breath' (or Spirit) of God. Moreover, that Adam was made alive by the breath of God implies that his life as this 'soul' was never independent of the will of God and the Spirit of

God (Gn. 6:3; Ec. 12:7; Mt. 10:28). The question of whether Adam was created mortal or immortal prior to the fall may miss the point by following Plato in presupposing some form of immortality that is independent of the will of God. Human life is never to be conceived of in terms of an independent immortality since that life is never independent of the will and Spirit of God . . . In consequence of his fall, death was pronounced as God's judgment upon Adam since the relationship which was the basis of his 'effective immortality' had been broken. It is this breach of spiritual relationship which constitutes the 'spiritual death' that characterizes the totality of human existence without Christ (Rom. 7:9; Eph. 2:1ff.) . . . As in the case of the biblical words traditionally translated 'soul' *(nepeš; psychē), the Hebrew and Greek words to express physical, emotional and psychological being are an interpreter's minefield (NDT* 29–30, italics added).

How much of a minefield becomes apparent when we examine how the translations have changed over the centuries.[1]

The human person as a living soul (nepeš and psychē)

Genesis 2:7 tells us that 'The LORD God formed man from the dust of the ground and breathed into his nostrils the breath of life, and the man became a living being.' The common assumption is that this tells us how man was made *in the image of God,* namely, by being given an immortal soul. It is then further boldly asserted that this is what distinguishes humankind from animals. Regarded in this way, this verse is seen to parallel Genesis 1:27, where we read:

> So God created man
> in his own image,
> in the image of God
> he created him;
> male and female
> he created them.

But it is important to note that the reference to man becoming a 'living being' in Genesis 2:7 is a translation of *nepeš,* which refers to non-human animals on four occasions in the first chapter (Gn. 1:20, 21, 24, 30). The straightforward interpretation seems to be that *both* humankind and animals *are* living beings or souls. They are not to be regarded as being made up of two kinds of 'stuff', a soul and a body, which can be separated and which subsist independently. According to the Bible, the 'soul' describes the *whole* of the living organism, whether animal or human, and comprises the body as well as mental powers. Living beings are spoken of as *having* soul, that is, conscious being, in order to distinguish them from inanimate objects that have no life. There are nineteen passages in the Old Testament and one in the New Testament which use the word *nepeš* or its Greek equivalent *psychē* in connection with animals.

On occasion, *nepeš* refers to a *dead body*; for example: 'He must not enter a place where there is a dead body' (*nepeš*; Lv. 21:11). The Bible also speaks of human death as *the death of the soul*; there are at least thirteen such instances in the Old Testament. In Scripture, a human being *is* a soul, in the same way that an animal *is* a soul. The difference between human and animal is not, in this sense, to do with soulishness.

More detailed study of *nepeš* in the Old Testament and of *psychē* in the New Testament shows a number of shades of meaning depending on the context in which the words are used. Before dealing with them, however, we should note that the world *psychē* was used in the Homeric poems in the eighth and seventh centuries BC to mean the 'life' of the whole person, as well as the seat of the desires and of thought; the *psychē* was consistently represented as surviving after death as a ghost in a shadowy world. In the fifth and fourth centuries BC, culminating with Plato, the idea of *the immortality of the soul* was elaborated. The association of *psychē* with *nepeš* led to its introduction into Christian thought around the end of the second century AD; only subsequently was the notion of immortality read into the word *psychē* in its New Testament occurrences. This background is important in understanding how particular meanings of 'soul' became incorporated into Christian thought.

1. *'Soul' meaning 'me' or 'myself'.* There are many uses of *nepeš* in the Old Testament and *psychē* in the New Testament where 'my soul' equals 'me', and 'his soul' equals 'him'. Thus the word may be used with a proper noun so that, for example, 'David's soul' would be equivalent to 'David' or 'David himself'. In this sense *nepeš* is used more than 280 times in the Old Testament. For example, in Genesis 27:19 the AV reads, 'Sit and eat of my venison, that thy *soul* may bless me', while the NIV reads, 'Sit up and eat some of my game so that *you* may give me your blessing.' In Leviticus 16:29, the AV reads, 'you shall afflict your *souls*', but the NIV reads, 'you may deny *yourselves*'. For Numbers 16:38 the AV reads, 'these sinners against their own *souls*', while the NIV has 'men who sinned at the cost of their *lives*'. In Deuteronomy 14:26 the AV has 'whatsoever thy *soul* lusteth after', and the NIV 'whatever *you* like'. In Isaiah 38:17, where the AV reads, 'Thou hast in love of my *soul* delivered it from the pit of corruption', the NIV renders, 'in your love you kept *me* from the pit of destruction'. Note incidentally that the death from disease that Hezekiah was expecting in this passage was the death of his *soul*. In the New Testament there are 24 examples of the use of *psychē*

equivalent to the Hebrew *nepeš* in this sense, seven of which are quotations from the Old Testament.

2. *'Soul' associated with the emotions.* There are more than 120 Old Testament passages where the soul (*nepeš*) is specially connected with the desires or emotions (for example, where AV reads 'my soul loveth') including one instance where the desire is that of an animal (Je. 2:24). In addition, there are 21 instances in the Old Testament where the word 'soul' is added to the word 'heart'; the consensus is that these mean 'with all [my] might and main'. In the New Testament the word *psychē* is used on 12 occasions in this sense. As with *nepeš* in the Old Testament, *psychē* indicates an inner aspect of a person, but there is no hint that the *psychē* alone carries personality, or that it survives the body, or that it is immortal.

There are more than 50 Old Testament references where mind and feelings are closely related. Examples include Genesis 23:8, where the *nepeš* is the organ of resolve; Exodus 23:9, where it speaks of feelings in general; Leviticus 26:16, where it is the seat of sorrow; Deuteronomy 23:24, where it is the seat of desire; Judges 18:25, where it is the seat of anger; 2 Kings 25:6, where it is the seat of joy; and Proverbs 27:9, where it is the origin of good counsel. Nowhere is there any justification for making such feelings into separate 'things'; rather, they focus on particular aspects of a unified person as required by the context of what is being discussed or described.

3. *'Soul' meaning 'life'.* There are more than 150 occurrences of this in the Old Testament. For example, Exodus 21:30 in the AV reads, 'he shall give for the ransom of his *life* whatsoever is laid upon him', while the NIV renders, 'if payment is demanded of him, he may redeem his *life* by paying whatever is demanded'. Isaiah 53:12 in the AV says, 'because he hath poured out his *soul* unto death', which becomes in the NIV 'because he poured out his *life* unto death'. The sense of 'life' is the most frequent use of *psychē* in the New Testament, there being some 46 instances. Here *psychē* is sometimes translated 'life', sometimes 'soul'. The basic meaning is always of the person or the self, as in Matthew 10:39, where we read, 'Whoever finds his life will lose it, and whoever loses his life for my sake will find it' (NIV).

Is the soul immortal?

There are a number of passages often quoted to argue for the separate existence and natural immortality of the soul.

First, there are passages where *psychē* in the New Testament is used to mean 'human being', in exactly the same sense as the Hebrew *nepeš* in the Old Testament. One passage in particular is often cited: Matthew 10:28 reads, 'Do not be afraid of those who kill the body but cannot kill the soul. Rather, be afraid of the One who can destroy both *soul* and *body* in hell.' At face value this could certainly provide a basis for the survival of a separate soul at death.

To 'kill the body' here, however, means to take the present life on earth. But this does not annihilate the person himself. It puts him to sleep in death. He is finally destroyed, according to other scriptures, in the second death, when his person or self is killed for ever.

In this regard we have to examine our Lord's declaration that Jairus's daughter was 'not dead but asleep' (Mt. 9:24). She was actually dead (equivalent to her body being killed), but as she was going to wake up she could rightly be said to be asleep (the word frequently used in Scripture to denote physical death). We return to the idea of death as sleep later in this chapter and ask what it may or may not imply.

Another seemingly confusing passage appears in Revelation 6:9, where *souls* are spoken of in a way which is often thought of as the disembodied spirits of the martyrs. In keeping with the whole of the Apocalypse, however, these verses are most naturally interpreted as symbolic. The key to their meaning lies in Leviticus 17:14, where the soul is identified with the blood. The souls are the dead persons of the martyrs.

A further passage often used to argue for the separate existence and natural immortality of the soul is Micah 6:7, where the prophet rehearses what the worshipper might render to God. Having listed various alternatives in verse 6, he continues, 'Will the Lord be pleased with thousands of rams, or with ten thousands of rivers of oil? shall I give my firstborn for my transgression, the fruit of my *body* for the sin of my *soul*?' (AV). This contrast between body and soul is retained in the NIV, which reads almost identically. The key is that the Hebrew word *bāśār*, 'flesh', is also frequently translated 'body'. In general it refers to separate parts of the body, while the body as a whole is often *nepeš*. Thus in Micah 6:7 the word translated 'body' in fact primarily means 'belly'; thus 'the sin of my soul' simply means 'my sin'. Hence the meaning becomes, 'Shall I offer my firstborn for my transgression, the fruit of my *belly* for my sin?' The REB renders the verse, 'Shall I offer my eldest son for my wrongdoing, *my child* for the sin I have committed?'

Another passage which has often confused people is 1 Kings 17:21–22, the account of the raising of the widow's son at Zarephath by Elijah the prophet. The AV has '. . . let this child's soul come into him again . . . and the soul of the child came into him again'. The NIV reads, ' ". . . let this boy's life return to him!" . . . and the boy's life returned to him'. This is a good example of the rightness of rendering *nepeš* as 'life' instead of 'soul'. The passage thus means, not that a separate part of the child went away and returned, but that the boy died and then came to life again.

A verse commonly used to support a tripartite model of humanness is 1 Thessalonians 5:23, rendered in the NIV 'May your whole spirit, soul and body be kept blameless at the coming of our Lord Jesus Christ'. The context is the key here. The apostle is exhorting his readers to a sanctified life, and thus he begins, 'May God himself, the God of peace, *sanctify you through and through.*' He wants his readers to be sanctified in every part; you name it, he wants it to be sanctified. He then re-emphasizes it when he goes on, 'May your whole spirit, soul and body be kept blameless . . .' Quite apart from the danger of building a whole doctrine on a single verse (particularly a doctrine inconsistent with the whole thrust of Scripture) we should note that it is Christians who are being addressed. Paul teaches elsewhere that Christians are only too aware of two aspects of their being, the Adamic nature which they owe to their birth and the spiritual nature which they owe to their rebirth. The former is called *soul*, and this is the *nepeš* of the Old Testament and the *psychē* of the New Testament; the *body* is the outward, visible form of this Adamic nature. In Thessalonians, the *psychē* is contrasted with the *spirit*, a translation of the word *pneuma* (see below). The same contrast is found in Hebrews 4:12, where we read, 'The word of God . . . penetrates even to dividing *soul* and *spirit*, joints and marrow . . .' Set in their proper context these words mean that the study of the Scriptures will indeed show us which desires, aspirations, emotions and thoughts lie on the old and sinful Adamic side of our nature, and which on the regenerate side resulting from the new birth. Similarly, Paul contrasts the 'natural man' and 'he who is spiritual' in 1 Corinthians 2:14–15, AV. Here 'natural man' translates *psychikos* (the adjectival noun related to *psychē*) and means someone who possesses *soul* but not *spirit* in the sense of a regenerate principle of life (or possibly 'the Spirit', as NIV).

Generalizing from the translations of *nepeš* and *psychē* as 'soul', we can say that normally they refer to the whole person; on occasions they point to particular aspects of the functioning of the individual which are characterized by linkage with, for example, emotions or feelings; but their most frequent use carries the meaning 'life'. There are more than 40 instances of this last use in the New Testament; sometimes the word is *soul*, but the basic meaning is that of the person or the self (*e.g.* Mt. 10:39). Above all, the key to a proper understanding of the varied usage of *nepeš* and *psychē* is the focus on the *whole* person rather than on a particular aspect of it.

The spiritual dimension of humankind

Just as *psychē* in the New Testament normally corresponds to *nepeš* in the Old Testament, *pneuma* in the New Testament corresponds to *rûah* in the Old Testament. Both words are usually translated 'spirit'. But does this imply that there is another part of human nature, and that 'spirit' is distinguishable from the rest of the human person and is indeed a separate part?

The basic meaning of *rûah* is 'wind'. As with *nepeš*, we find that both humans and animals possess it (*e.g.* Gn. 6:17; 7:15, 22; Ec. 3:21). It also appears in a very familiar passage in Psalm 31:5, where we read, 'Into your hands I commit my spirit.' These words were recalled by our Lord on the cross, and indicate that he entrusted to God his human spirit, the principle of life that he possessed as a human being, so that it could be restored to him in resurrection. In similar vein, Psalm 104:29–30 reads, 'When you take away their breath [*rûaḥ*] they die and return to the dust. When you send your Spirit [*rûaḥ*], they are created.' In Psalm 146:4 we read, 'When their spirit [*rûaḥ*] departs, they return to the ground'; that is, our spirit, the principle that makes us living beings and keeps us alive, is taken from us at death.

It is this same life principle which is referred to in the passage we have already quoted from Genesis 2:7. There we were told that the man was made of the dust of the ground and then that the 'breath of life' was given; it is a life principle issuing from the LORD God. It is used in this way at least 20 times in the Old Testament, referring to the life principle inbreathed by God; it is also noteworthy that in two passages, animals are included as well as humans (Gn. 7:22; Ps. 150:6. 'Breath' and 'spirit' in many of these passages are essentially parallels; see for example Job 34:14, 'he withdrew his spirit and breath'.

But there is another theme in the meaning and usage of the word *rûaḥ* in the Bible: the Hebrew Scriptures attach great importance and significance to an intangible quality which is captured in the *rûaḥ* of an individual. This quality is more than the sum total of our actions. While the word often denotes God's quality (the 'spirit' of God), it can be applied also to the non-believer, as Deuteronomy 2:30 and 2 Chronicles 21:16 (Revised Version) make clear. It can also be equated with character, in the modern sense of the word; that is, wisdom or folly, humility or pride. It is this spirit which shows what we are truly like and so, by implication, what the source of inspiration for living must be (*e.g.* Pr. 20:27).

The Old Testament knows nothing of a narrowly intellectual attitude to life. The Word of God speaks to and is written upon the heart; it is never given simply as academic information. From this it follows that scientific objectivity, which in our time we prize so highly, finds little place in Old Testament thought. Indeed, our reaction to events and circumstances of any kind, whether kindness or persecution shown by others, or around or to us, is never a dispassionate activity in the Old Testament Scriptures; it is always seen as a total involvement, whether it be anger or worship.

Corresponding to *rûaḥ* in the Old Testament is *pneuma* in the New. As well as referring to the life principle, *pneuma* is also often used simply to describe a person. There are more than 200 instances where it refers to the Holy Spirit and 56 instances where it denotes a person. Most of these, interestingly enough, are references to evil spirits in the synoptic gospels and the Acts of the Apostles. All three persons of the Trinity are at times referred to as spirit. Thus in John 4:24 God the Father is referred to as spirit ('God is spirit'); in

1 Corinthians 15:45 Christ is referred to as spirit ('the first man Adam became a living being [*psychē*]; the last Adam a life-giving spirit [*pneuma*].' And in other instances the references are clearly and less surprisingly to God the Holy Spirit. In all these cases *pneuma* probably means something like 'not existing in the way that fleshly, mortal beings exist'.

There are several places in the New Testament where the word *pneuma* carries the meaning of 'life principle' in the same way as *rûaḥ* does in the Old Testament. For example, Matthew 27:50 speaks of Jesus giving up his 'spirit', while James 2:26 reads, 'As the body without the spirit is dead, so faith without deeds is dead.'

There is yet another meaning of *pneuma* in the New Testament: on 29 occasions it refers to the regenerate nature. Here the sense of *pneuma* as a life principle is combined with its meaning as 'disposition' or 'character'. The new nature is certainly a new life principle; but it is essentially a moral life principle since it is a holy disposition or character.

As in the case of *nepeš* and *psychē*, so with *pneuma*, there are seemingly problematic proof-texts. One often quoted is 2 Corinthians 7:1: 'let us purify ourselves from everything that contaminates body and spirit, perfecting holiness out of reverence for God.' Here the apostle is exhorting believers to holiness and to the avoidance of all defilement in things of the flesh, by which he means immorality, and in the things of the spirit, by which he means false religion. There is certainly no need to reify the spirit on the basis of this text.

Of all the instances where *rûaḥ* and *pneuma* occur, only a handful can be construed as supporting the idea of the spirit surviving the body. These include Ecclesiastes 12:7, where the spirit is said, at death, to go back to God who gave it; and Luke 23:46 and Acts 7:59, where the spirit at death is commended into the hands of God. When these texts are set alongside all the others which show that the spirit is a life principle breathed into us to make us alive and conscious, however, it is clear that there is no warrant whatever for supposing it to be conscious when it returns to God; its 'return' simply means that God takes a person's life away.

There is one more word that appears in reference to the nature of humankind, namely the *heart*. The Hebrew and Greek words translated 'heart' are not used in Scripture in connection with the creation of humankind as are *nepeš* and *rûaḥ*. There are, however, some passages which seem to imply a possible separate existence for the heart. For example, in Psalm 22:26 the AV has 'Your heart shall live for ever'. The NIV has 'may your hearts live for ever!' While some may wish to argue that these are grounds for some sort of proof of universal immortality, one should note that the passage in question is concerned with the 'meek' and with 'they who seek the LORD'. The expression 'your heart' is commonly used in the Psalms to mean simply 'you' in the sense of the deepest aspect of our human function.

In the past it might have been described as the seat of the will and conscience. Certainly at times the heart means simply the person himself, as in the above case and in 1 Thessalonians 3:13, where we read, 'May he strengthen your hearts so that you will be blameless and holy in the presence of our God . . .' The meaning here is simply 'may he strengthen *you* so that you will be blameless . . .' In the Old Testament both 'heart' and 'kidneys' (AV 'reins') are used for the inner person, the seat of emotions, thoughts and so on – that is, mental life which was associated not with the brain but with the inner parts of the body.

Death

It would be easy to conclude from the biblical emphasis on the holistic view of the person that people cease to exist at death. Indeed, what we have written elsewhere (chapter 4) about the analogy of the computer with the human individual reinforces this. We suggested that when a mathematician talks about his computer 'solving an equation', he means that the behaviour of the machine in some sense currently embodies that particular equation. If that particular computer was destroyed, the mathematician, knowing the equation he was seeking to solve, could re-embody it in another different computer. In like manner, while physical death is for us the end of our mental and spiritual life here and now, that does not mean that God, if he so wished, could not re-embody it in a resurrection body. It is indeed in him that we live and move and have our being. While arguments of this kind from analogy have to be handled with extreme care, there is a real biblical sense in which death is our end in *this* space-time framework. Like our mathematician, God could, if he wished, set up the same equation in another embodiment. We shall return to this, but first we should note several key biblical themes concerning death.

The biblical picture of humankind is of a soul whose life is maintained by the spirit of life breathed into him at creation by God, and the direct cause of death is the removal of this spirit by God. The words 'to die', 'death', 'dead', in Hebrew and Greek, as in English, refer simply to the extinction of life. This is often conveyed by 'sleep', a way of thinking about death which appears throughout Scripture, as well as an everyday euphemism. For example, the result of the departure from a person of the life principle or spirit and its return to God is described as a state of sleep, in which there is no resemblance and no possibility of praising God (Ec. 12:7). In Hebrew, we meet the frequently occurring expression that So-and-so 'slept with his fathers'. The Hebrew word used is literally 'lay down', but in Acts 13:36 it is rendered by the Greek word which means 'to sleep'; thus the kings and others who died are said to sleep with their fathers. Thus our state in death would be final were it not for the resurrection of both the just and the unjust, which makes death temporary and turns it into a sleep. The dead, it seems, do not awake and are not raised from sleep until the end of the world.

In the New Testament, the same theme recurs. Thus in Matthew 27:52 we read that many bodies of holy people who slept (NIV translates 'had died') arose; in John 11:11 we read, 'Our friend Lazarus has fallen asleep', our Lord's own words; in Acts 7:60 we read, 'When he had said this, he fell asleep'; in Acts 13:36 we read 'David . . . fell asleep.' This is a quotation from 1 Kings 2:10, where the NIV translates 'David rested with his fathers.' Again, in 1 Corinthians 7:39 we read, 'A woman is bound to her husband as long as he lives. But if her husband dies, she is free to marry anyone she wishes.' It is noteworthy that the Greek word translated 'dies' is in fact 'should fall asleep'. Thus 'sleep' is here contrasted with physical life. We believe that this description of death as sleep justifies believers' assurance that the moment after they draw their last breath and close their eyes, they open them again in the presence of Jesus in resurrection glory with all their loved ones and the whole loving community of the church of God around them. They now have their resurrection bodies, their house not made with hands, eternal in the heavens. The apostle Paul's language in 2 Corinthians 5:6–8 is consistent with the subjective experience of a believer in passing instantly from this world to resurrection glory. So profound is his unconsciousness in death that after closing his eyes he opens them at what to him is the next instant on the resurrection morning. This was an important element of Paul's hope indicated in Phillipians 1:20–24.

With these thoughts in mind, it is natural to turn to 1 Corinthians 15, which contains the most explicit teaching on the Christian hope of eternal life. Here there is no suggestion of a physical continuity between our present body and the resurrection body. We are told that just as the plant that rises from the ground is very different from the grain that was sown, so at death we are 'sown a natural body . . . [but] raised a spiritual body' (1 Cor. 15:44). This continuity is essential at the level of our personal relationship with God; our personality will be re-expressed in this new embodiment, with the same essential relational structure that identifies and distinguishes us as individuals here and now. It is not essentially different from undergoing all kinds of metabolic change in our normal night's sleep, but waking up to find ourselves the same individuals who went to sleep the night before. The whole idea of resurrection remains a profound mystery, but there seems to be no basic logical difference for personal identity between waking up in another world and waking up in this world. As Donald MacKay has pointed out, both events depend utterly on the fiat of our Creator, to whom we owe our continuing identity moment by moment, day by day, here and now. If God chooses to know and recognize us in the resurrection as those whom he knew in the days of our flesh, then by the same token that is who we are; it is the Creator alone who determines and gives being to what is the case. 'In him we live and move and have our being' (Acts 17:28). Mysterious, to be sure, but sufficient for us at least to see through the glass darkly.

An intermediate state?

In the previous section, we stated that 'We believe that this description of death as sleep justifies believers' assurance that the moment after they draw their last breath and close their eyes, they open them again in the presence of Jesus in resurrection glory, with all their loved ones and the whole loving community of the church of God around them.' We recognize that this may, to some, be an overbold statement, since there is a tradition of a belief in some form of intermediate state. We believe that differences in views on this issue are, at the end of the day, a matter of hermeneutics. It is helpful, we believe, in discussing this topic, to identify common ground. First, those who are equally competent biblical scholars and share a common view of Scripture, while holding differing views on the intermediate state, emphatically share the central Christian hope, which is that we look forward to resurrection, to being with Christ, which is 'better by far' (Phil. 1:23). Secondly, any discussion of this topic should be set in the wider context of the biblical portraits of human nature traced out in our earlier discussions, which recognize that 'the traditions informing the New Testament writers are more variegated than normally thought . . . and that this underscores both the importance of attending to New Testament voices set within their own contexts and sometimes *ad hoc* modes of argumentation and of identifying where the New Testament writers have articulated with and/or over against other cultural voices' (Green, forthcoming). Professor Green suggests that

> 1) the New Testament is not as dualistic as the traditions of Christian theology and biblical interpretation have taught us to think, though enough conceptual glossalalia exists amongst New Testament witnesses for us to see how a dualist reading of human nature has developed among Christians; 2) nonetheless, the dominant view of the human person in the New Testament is that of ontological monism, with such notions as 'escape from the body' or 'disembodied soul' falling outside the parameters of New Testament thought; 3) New Testament writers insist on the concept of soteriological wholism and, in their portraits of human nature, place a premium on one's relatedness to God and to others; and 4) the emphasis on anthropological monism in the New Testament underscores the cosmic repercussions of reconciliation, highlighting the notion that the fate of the human family cannot be dissociated from that of the cosmos.

Professor Green then comments on one passage regarded by some as arguing for an intermediate state:

> Undoubtedly, the most pressing evidence in Paul for a body–soul dualism is found in 2 Corinthians 5:1–10, and especially verses 2 and 3, translated in the NRSV as follows: 'For in this tent we groan, longing to be clothed with our heavenly dwelling – if indeed, when we have taken it off we will not be found naked.' According to many interpreters, Paul presents here a thanatology concerned with the freeing of the soul from the body for a higher destiny. As one recent study puts it, 'Paul speaks of three states: the present condition in the tent-like frame, the intermediate state of nakedness, which he does not find desirable, and the future condition in which a further frame will have been put on, hopefully, over the present one'. 'The earthly body is a tent-like existence, a temporary shelter. Paul seems to believe that believers' resurrection bodies are already prepared in heaven, in heavenly cold storage, so to speak' [Witherington 1995: 391]. This is a possible reading, but it is not the only one. In fact, if Paul is concerned here with thanatology, which this view requires, then he has digressed rather dramatically from the focus of this larger section on apostolic suffering (4:7–5:10). Moreover, it involves Paul in a discussion, the focus of which would be uncharacteristically subjective and individualistic.
>
> Victor Furnish [1984: 292–295] has offered an alternative reading that takes with greater seriousness the place of this passage in Paul's larger argument. In this case, what is on centre stage is the frailty of human existence, and the concomitant possibility of denying that one has been clothed in Christ if one suffers as Paul has suffered. Paul, then, would be speaking of his having been clothed with Christ at his baptism, a well-known metaphor, *and* longing for the completion of his salvation (*i.e.* his being 'clothed over' and therefore not found naked in the final judgment). As is typical of Pauline thought, the duality here would be eschatological, focused on the tension between the now and not yet, not anthropological.

Another line of biblical evidence regarded by some as pointing to an intermediate state points to references in the Old Testament to 'shades'. Thus, some biblical scholars take the view that those who lived in Old Testament times believed that the dead are 'shades' (*rᵉpa'îm*, as in Psalm 88:10), which was the commonly held ancient view, like that of the Greeks before Plato. A 'shade' represents a sort of shadowy version of the bodily person in a dark and gloomy underworld. These scholars point to the word *pneuma*, which is used in this sense in Luke 24:37, 39. Thus the *rᵉpā'îm* are thought of as shades or ghosts rather in the Homeric sense. Others argue that such an idea never occurs elsewhere in Scripture, and suggest that the idea may well have arisen from the poetic imagery in which the word occurs in Isaiah 14:9. It may indeed be that there was such a belief in ancient Israel and Judah, but such a false belief would never be considered as normative teaching. In Isaiah 14:9 and 26:14, the word refers to dead kings or lords of the past; in Isaiah 26:19, the *rᵉpā'îm* appear to be contrasted to the blessed dead. In Proverbs 2:18; 9:18 and 21:16, the word seems to imply the dead in general. There does not seem to be anything in these occurrences that obliges us to put the meaning 'shades' upon the word. It seems unreasonable to insist upon it.

This line of argument may be developed by references in the New Testament to, for example, ghosts. See, for example, Luke 24:37–39, where the apostles, on seeing the Lord, believe they are seeing a ghost. Again, this, as in the case of the *rᵉpā'îm* of the Old Testament, may have been a commonly held view in that culture at that time, which was therefore naturally and understandably absorbed into their thinking and was consequently a perfectly natural response to their unusual experiences. That, however, does not necessarily imply that that and similar passages are teaching us to believe in ghosts. In and of themselves they would be insufficient as firm grounds for arguing for some form of intermediate state.

Finally, some of those who argue for an intermediate state point to references in Scripture to paradise. We may ask, therefore, what about the New Testament teaching where the world to come is twice referred to as 'paradise' (Lk. 23:43; Rev. 2:7)? Is it natural to think of this as an intermediate state? The word translated 'paradise' is borrowed from the Persian, meaning an orchard or fruit garden, and implies the restoration of Eden together with the innocence and happiness that Adam and Eve enjoyed there. The first of the two occurrences is the Lord's words to the dying thief on the cross: 'I tell you the truth, today you will be with me in paradise' (Lk. 23:43). They suggest the survival both of the Lord Jesus and of the thief in a disembodied state after their death, and their presence together in paradise on that day. It is in this sense that they are normally taken. This, however, contradicts the thrust elsewhere in Scripture, and some Greek scholars argue that the phrase can be better translated as 'Truly I say to you today, you will be with me in paradise.' They point out that the expression 'I say to you today' is an accustomed phrase for emphasis in Hebrew. Thus we often find Moses speaking of the commands 'which I am giving you today' (Dt. 4:40; *et passim*). Moreover, the day on which the Lord spoke to the thief was the very day which made the thief's entry into paradise possible by Christ's own suffering and death, which was taking place upon it. Thus the Lord's answer was an exact response to the poor thief's request that he would remember him when he came into his kingdom (verse 42). Scholars who take this view believe it is strengthened when notice is taken of the second occurrence of the word, in Revelation 2:7: 'To him who overcomes, I will give the right to eat from the tree of life, which is in the paradise of God.' But note that the tree of life is on the *new* earth (Rev. 21:1; 22:2), which does not come into being until the day of judgment is finished and this world is purged by fire (2 Pet. 3:13). To base an argument for an intermediate state on two references to paradise, which are capable of an interpretation in line with the rest of Scripture on this matter, is very precarious.

It would be inappropriate for us to go beyond noting these differing views of biblical scholars, other than commenting that the view which finds the evidence for a clear teaching about an intermediate state in Scripture is, on the face of it, to us, less congruent with our overall view of human nature developed from within science than is the alternative view as put forward by Joel Green.

There is one footnote to all of this which we believe is important. In any discussions such as those in the preceding paragraphs, it is crucial to remember that we are dealing with matters which are ultimately a mystery. For the person of faith, the key point to note is that in the face of the mystery of death, sufficient clues have been given for our comfort. It is clear that we shall not be disembodied; we shall be 'clothed'; we shall have a glorious body, and we shall indeed be 'spiritual bodies' (1 Cor. 15:44). Above all, it is important to note the emphasis of the apostle Paul: when writing to the Christians at Philippi, his strong affirmation is that 'to me, to live is Christ and to die is gain . . . I am torn between the two: I desire to depart and be with Christ, which is better by far' (Phil. 1:21–23). Thus it is clear that in his way of thinking, to die is to be with Christ.[2]

Resurrection

The Apostles' and Nicene Creeds concern themselves not with natural immortality nor with survival, but with resurrection. They speak of 'The Resurrection of the body, And the life everlasting' and 'the Resurrection of the dead, And the life of the world to come'. Resurrection is clearly taught and foreshadowed in the Old Testament (*e.g.* Is. 26:19; Ezk. 37:1–14; Pss. 16:10–11; 17:15; Jb. 14:14–15; 19:25–27). As we move on to the New Testament, it is soon evident that the model for the coming resurrection of the people of God is to be that of the Lord Jesus Christ himself. The resurrection appearances of the Lord demonstrated two principles: the identity of personhood, and a change of nature. The direct teaching of our Lord about the resurrection comes in his answer to the Sadducees who denied it (Mt. 22:22–33; Mk. 12:18–27; Lk. 20:27–40). In the gospel of John, our Lord gives four clear promises of resurrection. First, the work of raising the dead and making them alive is that of both the Father and the Son (Jn. 5:21). Secondly, all who are in the tombs will hear the voice of the one who is both Son of God and Son of Man and will 'come out – those who have done good will rise to live, and those who have done evil will rise to be condemned' (Jn. 5:28). Thirdly, the Lord Jesus will not lose a single one of his believing people, because it is the Father's will that everyone who believes on the Son should have everlasting life, and the Lord Jesus 'will raise him up at the last day' (Jn. 6:40). Fourthly, there is the marvellous and well-known promise of our Lord at the grave of Lazarus: 'I am the resurrection and the life. He who believes in me will live, even though he dies; and whoever lives and believes in me will never die' (Jn. 11:25–26). The clear teaching here is that resurrection and everlasting life are the gift of Jesus alone, and that the believer will be raised to life even if he dies.

As we move on to the teaching of the apostles, once again the key passage is 1 Corinthians 15. In this, the apostle Paul refers to some of the 500 believers who had seen the Lord and had 'fallen asleep', a picture we noted earlier as a significant in Scripture. The teaching is clear that it is because Christ has risen that the believer has any hope of resurrection. From verse 35 onwards, Paul moves towards a grand climax. Raising the question of the method of resurrection, he compares death and resurrection (as we saw earlier) to the sowing of seed in the ground and the appearance of the plant when it comes up. The two are utterly unlike, yet an identity runs through them. The bodies of those who rise differ, just as earthly creatures do among themselves. The body is sown in weakness, but raised in power; it is sown a natural (*psychikos*) body, but is raised a spiritual body. This agrees with the fact that the first human being Adam was made a living soul (*psychē*) and the last Adam a life-giving spirit (*pneuma*). Paul teaches that as we have borne the image of the earthly, so we shall bear the image of the heavenly. Some have argued from 2 Corinthians that to speak of the body as a building or a garment implies a spirit or person that continues to live separately from it. This need mean no more, however, than that there is a mind within the body and joined to it, and in view of the direct scriptural teaching we have reviewed above, it would be mistaken to interpret it otherwise. The human being is indeed (as we have argued throughout) what today would be called a psychosomatic unity.

Conclusions

The reason for this biblical digression is the renewed interest in human nature stimulated by scientific advances in psychology, neuroscience, ethology, genetics and cognate disciplines. Each of these sciences, with their specialized tools and possibilities for different levels of analysis, continues to shed new light on what it means to be human.

Biblical understanding has also changed. It is important to emphasize this. The word commonly translated 'soul' in the older English versions has more or less disappeared in recent translations. Quite often it has become 'I myself' or 'he himself', a kind of usage we meet frequently in the Psalms. The traditional assumption that the soul is some distinct entity which is somehow built into, or constitutes a separate part of, the human person is not scriptural. We have noted a fascinating parallel in the way that the 'coming into existence' of living beings, referred to in the Bible as being 'God-breathed' or as the conferring of a 'life principle', is reflected in what happens at death, when the 'life principle' or 'spirit' is taken from the organism and returns to God. This applies equally to animals and to humans.

Debates about the issues raised in this chapter will continue. Our concern is to help Christians faced with the weight of evidence of the ever-tightening link between mind, brain and behaviour. We are impressed by the way biblical scholars have been at pains to warn us away from earlier interpretations of Scripture which tended to reify the soul, the spirit and other aspects of personhood. It is important to look afresh at the biblical evidence and see how it produces portraits of human nature complementary to, and not inevitably conflicting with, the emerging scientific picture.

Hellenistic inputs have had a powerful and valuable impact on the development of Christian theology. But we have to be aware of such influences, particularly when we consider concepts like the soul. At any particular epoch in the history of the Christian church, the prevailing theological climate may affect the translations of the Scriptures made at that time. The Authorized Version of the Bible is no exception to such influences.

What we are concerned to stress is that the central and dominant emphasis in Scripture is the unity of individual human beings. Human persons are functioning beings who manifest several different aspects of their nature as they live and act. The Bible uses different words to describe function, such as the 'heart'. What distinguishes us from other living creatures is the clear teaching that in some profound sense we are made in the image of God. This implies neither that we are immortal nor that we have an immortal soul. There is no more warrant for this than there is for assuming that we share with God his omniscience or omnipotence. The distinctive teaching of Genesis, reiterated in various ways throughout the Bible, is that our role (and nature) is to care for God's creation: to subdue it, to rule over the fish of the sea and the birds of the air and over every living creature that moves on the ground, exercising the delegated authority of a responsible steward. In this, we are accountable to God (p. 222). The same theme recurs in Psalm 8 in the form of the question 'What is man?' Elsewhere in the Old Testament the question is asked in various forms, but is ultimately answered only in the New Testament with reference to Christ. It is Christ who is 'made a little lower than the angels' and it is he who is 'now crowned with glory and honour because he suffered death' (Heb. 2:6–9). There can be no truly authentic knowledge of human nature independent of what is disclosed to us in Christ, the perfect human being. Karl Barth argued that the person of Jesus Christ alone is the determinative source of any valid theological anthropology, and spoke of Jesus as the revelation both of the real human being which we are and of the true human being which we are not. The true goal and nature of human life can be discerned primarily by studying him and only secondarily by looking inwardly at ourselves. We agree wholeheartedly.

The precise identity of the 'image' in describing humankind as made in the image of God continues to be an issue of lively debate in Christian thought and doctrine. As far as the 'soul' is concerned, the best idea of 'the image of God' should not be thought of in static or individualistic terms but rather in functional terms, in terms of relatedness. Thus men and women are called in Christ *to be* the image of the eternal in the relatedness of the Trinity (Jn. 17:21–23).

The Bible makes no claim that the Christian's hope for the future is bound up with any idea of possessing a natural immortal soul; on the contrary, its firm assertion is that our hope is fully embedded in the resurrection. The final resurrection will be a divine creative act just as much as our original coming into existence was also due to God's gracious creative act; 1 Corinthians 15:44 indicates a clear discontinuity between our 'natural' and 'spiritual' bodies. The great creeds of the Christian church all emphasize the doctrine of the resurrection and say nothing at all about any doctrine of natural immortality.

The Hebrew-Christian view of human beings is a message for us about our calling, our nature and our destiny. We are called to worship and honour our creator, to exercise a stewardship over the creation as an obedient child to a father, and to enjoy fellowship with our Father/creator while standing in awe of him as creature to creator. We are encouraged to recognize the many-sidedness of his mysterious nature. Consequently, we must hold in balance the three aspects of our nature highlighted by Old and New Testament writers alike: physical make-up; capacity for mental life; and ability to make moral decisions, including an appreciation of the importance of a spiritual dimension to life. Working harmoniously together, these lead to the maintenance of a right relationship with God and with other human beings, a state which the Bible calls 'salvation'.

Chapter 9

Human nature: biology and beginning

We are more than animals, and we are not all the same biologically (never mind our spiritual state). Moreover, we are much more than the product of genes laid down at conception. The understanding of humanness derived from Bible and biology taken together enables us to consider reproductive 'technology' as part of our responsibility for the maintenance of God's work in a fallen world, rather than interference with innate sacredness or givenness.

We are biological machines. But Christians believe that we are also spiritual beings, made in the image of God. The difficulty is that a biologist could describe in principle all the genetic, embryological and behavioural processes that lead to the formation of an individual, without needing to refer to God once. Is there then room for a soul? Is 'humankind made in the image of God' wishful thinking?

We have described the essential basis of our humanness in chapter 8, locating the distinction between us and the animals in our possessing the 'image of God'; non-human animals do not have God's image. This does not affect the fact that animals are just as much a part of God's creation as we are. As we shall see in chapter 12, we have a special duty of care for animals because God has created them and entrusted them to us.

Charles Darwin was well aware that the most radical impact of his ideas would be in their implications for humans. In *The Origin of Species* he tried to divert attention from this. He mentions the evolution of humankind only once 'In the future I see open fields for far more important researches . . . Much light will be thrown on the origin of man and his history.'

Early and significant support for Darwin came from Charles Lyell. Ever since his own work on geological change had first opened up the question of the origin of species, Lyell had been disturbed at the prospect of degrading humankind through a link with animals. But in *The Antiquity of Man*, Lyell accepted the archaeological evidence for the existence of primitive 'sub-humans' and had to face the prospect of human evolution. He admitted the plausibility of Darwin's conclusions, and was even prepared to accept a gradual evolutionary progress through history – something he had been arguing against for decades. Where he baulked was at the appearance of humankind as a continuous development from its closest animal relatives; he believed that the distinctive qualities of humans must have been produced by a sudden leap in organization, taking life to a new and higher plane. Alfred Russel Wallace, co-discoverer of natural selection with Darwin, came to a similar conclusion. It led him to a belief that we possess a soul capable of existing independently of the body, and he became a spiritualist. According to his 1884 Bampton Lectures, Frederick Temple (who became Archbishop of Canterbury in 1896) believed that the 'enormous gap' separating human from animal nature might have been due to a 'spiritual faculty . . . implanted by a direct creative act' (1885: 186).

Ironically, these early post-*Origin* interpretations of man were obscured by the scientific arguments about natural selection and the mechanism of evolution, and have been submerged by the speculations of Teilhard de Chardin, Julian Huxley, and others.

Human variation

The Bible is absolutely clear that every human being is unique and knowable by God as an individual. In the parables of the sower and of the talents, Christ spoke about everyone having different gifts (Mt. 13:23; 25:5), and both Paul and Peter emphasize the need to use our own particular gifts if we are to fulfil the role that God has for us (Rom. 12:6; 1 Cor. 7:7; 12:4–11; 1 Pet. 4:10–11; *etc*.). God recognizes and cares for each one of us personally, and not merely because we are members of a particular category, such as his children or those who acknowledge him (Jn. 10:3; *etc*).

Running alongside this teaching on divinely determined uniqueness is a dominant theme in Scripture: that we are all alike created by God's sovereign grace, although we may be redeemed and sanctified; there is no possibility of living the life God has purposed for us except through his way. Christian teaching emphasizes the need for personal regeneration and spiritual growth, and stresses the impossibility of achieving these unless we pass through the one 'narrow gate' of which Christ spoke (Mt. 7:13).

This leads to the idea that there exists such a thing as a *normal* Christian life: conformity is identified with sound doctrine, and any deviation is heresy. There is a widespread image of a Christian as uniformly subfusc – drear, earnest in conversation and conservative in habit. This image is changing, partly on theological grounds and partly on sociological ones (changes in worship patterns and so on), but the goal to be 'like Christ' remains normative (2 Cor. 3:18).

How then do we reconcile the immensity of human variation with the call to live in conformity to the divine image? When Paul tells us that 'There is neither Jew nor Greek, slave nor free, male nor female, for you are all one in Christ Jesus' (Gal. 3:28), does this mean that we are to eliminate our differences, and that failure to do so is sin? Put bluntly, are all differences between human beings the result of the fall and therefore eradicable by grace, or is human variety God-given, in which case its suppression is wrong? Sexual difference antedates the fall, and is described by God as being 'very good' (Gn. 1:27, 31; 2.20–23), but a traditional interpretation of the story of the tower of Babel (Gn. 11:1–9) is that racial and linguistic differences stem from the Babel fiasco; in other words, humankind was culturally homogeneous until that post-fall event.

In contrast, the trend in secular western societies is to emphasize the importance and power of individual traits. To live a full life, we must express our individuality. In particular, if our personal idiosyncrasies are inherited, the individual is applauded for building on them, and excused from moral blame if they are antisocial (even if some expressions of personality have to be restrained for the good of society). Liberal culture has its problems here: although it is prepared to allow and encourage musicians to be musicians, mathematicians to be mathematicians, and homosexuals to be homosexuals, even it draws back from permitting the aggressive to be muggers or kleptomaniacs to be licensed robbers.

Variation can be inherited or environmentally acquired, but even environmental variation depends on the underlying genetical reaction system, dependent on the twenty-three chromosomes supplied by our father, and the twenty-three by our mother. If the fusion between two particular paternal and maternal gametes had not occurred, we would not exist. Those forty-six threads of a simple and chemically understood complex between protein and deoxyribonucleic acid (DNA) are the ultimate biological basis of each of us. It is wholly incorrect, however, to claim that the 'genetic code' of the 20,000 or so genes on the forty-six chromosomes gives a *complete or adequate* specification of the human being that develops from the fusion of egg and sperm. We need to recognize at least three factors which destroy the facile assumption of genetic determinism.

Genetics and epigenetics

The DNA of the chromosomes is 'translated' into chains of amino acids (polypeptide chains), which in turn form proteins and enzymes. But the production of polypeptide chains is not an automatic production process that goes on throughout life. Genes are subject to precise control or regulation, with many (indeed, most) being switched off most of the time. The genes in every cell in the body are affected by the history and environment of that cell, and the bulk of the chromosome set which is replicated in virtually every body cell is non-functional. Cells in which this control process breaks down are liable (if they survive) to be cancer-producing.

The development and functioning of a whole organism, however, is even more complicated. Some of the proteins that are primary gene products are recognizable in a normal body; they turn up as enzymes controlling vital processes or antigens affecting particular immunological reactions. The products of most genes interact together in the body to form secondary compounds, which are building-blocks for growth and maintenance, hormones, and so on. These interactions are highly complex and specific; in no way can the human body be properly seen as the simple consequence of a set of random chemical specifications. Although virtually all our characters can be regarded as affected by genes, their inheritance should be described as *epigenetic* rather than genetic; in other words, it is the result of processes acting subsequent to the primary action of the genes themselves.

Genes and environment

Primary gene products are the direct consequence of a rather simple chemical process that has been worked out in the revolution of molecular biology that began in 1953 with James Watson and Francis Crick's elucidation of the structure of DNA. At this level, inherited characters can be said to be determined by the genes carried by an individual. Once we leave the primary gene product level, however, the occurrence, speed and direction of the chemical processes in the body are affected to varying extents by environmental influences, inside as well as outside the body. This is of considerable importance in clinical medicine, because inherited defects in metabolism can often be corrected once they have been identified. For example, diabetes can be treated with insulin, haemophilia with anti-haemophilic globulin, phenylketonuria and galactosaemia by withholding from the diet phenylalanine and galactose respectively. It is not true that genetic disease cannot be treated, as used to be believed.

The interaction of genes and environment applies throughout normal development. Prenatal growth is slowed by the mother smoking, and maternal drinking may reduce the intelligence and size of a baby at birth ('foetal alcohol syndrome'). Childhood growth can be affected by nutrition. IQ is higher in first children, and in small families than in larger ones.

It is often difficult to work out the details of gene–environment interactions in humans where experimental breeding and environmental control cannot be carried out. Comparison of the behaviour and achievements of identical and non-identical twins, and of adopted and natural children, can go some way to understanding these processes, but unchallenged conclusions are very limited. Criminality and sexual deviation have often been attributed to family or to inherited influences, but the grounds for distinguishing between these are usually equivocal. Notwithstanding, there can be no doubt at all that we are affected radically by our environments as well as by our genes.

For the moment all that is necessary to note is that few behavioural traits are irrevocably determined by genes; virtually all human behaviour (and associated characters, like intelligence) can be unintentionally or consciously influenced by the environment.

Although identical twins (that is, genetically identical individuals) are often strikingly alike in both behaviour and physical traits even when reared apart, nevertheless there are plenty of examples where identical twin pairs show marked and significant differences for some characters. Our genes cannot express themselves in a vacuum; even in cases where our genes predispose us towards certain characteristics (as a shallow hip-joint predisposes to congenital hip dislocation, or an extra Y chromosome pushes us in the direction of mindless aggression), there is no *irrevocable* association between genes and the physical or behavioural trait that finally emerges.

Foetal death

The third factor is different, because it affects both individuals and changes in the genetic make-up of a population between conception and birth. So far, we have considered the complex relation between genes and characters. There is also a non-simple relationship between the fertilized egg and survival to adult life. For every 100 eggs subject to normal internal fertilization, 85 will be fertilized if intercourse is frequent, 69 will implant, 42 will be alive one week later, 37 at the sixth week of gestation, and 31 at birth. Between a third and a half of the foetuses that spontaneously abort in the first few weeks of pregnancy have an abnormal chromosomal complement: 97% of babies with a single sex chromosome (Turner's syndrome) and 65–70% with Down's syndrome (mongolism) miscarry by the eighteenth week. It seems likely that a large proportion of foetuses with anomalies of the central nervous system (such as anencephaly and spina bifida) also miscarry spontaneously. Survival to birth is not the norm; it occurs in only a minority of conceptuses, and many of those eliminated are recognizably abnormal in their genetic complement.

When we consider together interactions of genes with genes, interactions between genes and the environment, and foetal death, it is obvious that we are determined by our genes only in a very loose sense. Some of the differences between us are the result of differing genetic variants, but these differences can be magnified or diminished by family, social, educational, cultural, or other environmental influences. Even someone whose hereditary make-up irrevocably fixes some traits – such as a person with Down's syndrome – nevertheless can show a wide range of behavioural responses from gross mental defect to near normality. Although in one sense we must be regarded as the sum of the genes we acquire at conception, in another sense we are considerably more than that sum. Just as the wetness of water is a property that cannot be predicted from the atomic properties of its constituent hydrogen and oxygen, so human beings cannot adequately be described by their genes, even if we could know the details of the entire DNA code of an individual.

Life – and death

Understanding our biological nature is essential to understanding our nature as unitary beings, made in God's image. Although we are animals, we are not simply animals with divine essence added (and, by implication, with the possibility of its being subtracted). But having said this, we must remind ourselves that the creation of humankind is described in the Bible as the making of the physical form by God, who *then* 'breathed into his nostrils the breath of life, and the man became a living being' (Gn. 2:7). It would be wrong to assume from this that body and soul are distinct; as we have seen, the Hebrews did not have separate concepts of body and soul (pp. 140ff.). Dualism is conventionally regarded as simply a Cartesian materializing of body and spirit (or soul), but there are different sorts of dualism. For example, Paul contrasts two human 'natures' characterized respectively by disobedience and filial obedience. This latter dualism resonates with the interpretation given above, which sees the image of God in humankind as about relationships, with the fall primarily representing a break in the relationships of God, humankind and nature (see p. 155).

This gets us close to our humanness in the fullest sense, but, to understand it fully, we have to dispose of some very widespread misconceptions about the nature of human life and our origin as individuals. The role of fusion of male and female gametes in forming new human potential was not recognized until the seventeenth century; the cellular details of this were not described until the end of the nineteenth century. To the Bible writers (and for many centuries after them), 'life' was a great mystery. *Cogito ergo sum* ('I think, therefore I am') was as near as secular man could get to describing it. We now know about physical life in great detail. Indeed, some laboratory

experiments have come close to making 'new' life.
We understand the molecular biology of human reproduction in great detail. At that level life is continuous: gametogenesis, fertilization, embryogenesis, birth, growth, reproduction, fertilization, can be divided into recognizable periods, but the divisions are arbitrary. Furthermore, the gametes are just as unique as the fertilized ovum; the embryo is unable to live independently of its mother for at least four or five months after fertilization (and nowadays that stage is more dependent on medical technology than on biology); and so on. Although we can speak perfectly correctly of a developing baby as either 'a miracle of new life' or as merely 'another manifestation of continuing human existence', we are using different language for the same reality (or describing 'different levels') when we do this. It is particularly important to recognize this distinction when we are speaking about 'life'. Sometimes we are referring only to 'biological' processes, but at other times to 'human' life in the biblical sense of 'made in the image of God'.

Reproductive developments

This leads us to consider the reproductive process itself. Children are explicitly described in the Bible as gifts from God (Pss. 127:3–4; 128:3); theologically, they are one of the purposes or 'goods' of marriage (Augustine listed the three 'goods' as offspring, fidelity and sacrament; in modern terms, they are commonly referred to as procreational, moral and relational). On these grounds, some have condemned scientific or medical intervention in reproduction as contravening God's natural laws. Such commentators distinguish between the natural begetting and artificial making of children; intervention is condemned as 'playing God'. In some ways, the distinction between natural and artificial is the same as the one between mechanism and purpose, which (as we have repeatedly seen) describes complementary causes rather than alternatives. Reproduction, however, is concerned with life and its control; it would be inappropriate to ignore the morality of developments in biomedical science.

Christian assumptions

It is frequently asserted as almost self-evident that 'life begins at conception'. In fact, this is a proposition that became generally accepted only in the mid-nineteenth century, formalized by Pope Pius IX in a papal bull (*Apostolicae Sedis*, 1869) because it was an implication of the 'immaculate conception of the Blessed Virgin Mary'. Prior to this time, the 'Christian tradition' (as set out in canon law and in English common law, which prescribed the penalties for carrying out an abortion) was that killing a foetus in the first two or three months of gestation was not punishable as homicide, albeit still regarded as a serious misdemeanour and condemned by the church.[1]

We can recognize four main themes in biblical teaching.

1. *Prenatal life*. There are many references in the Bible to life before birth. The Psalms often refer to God's care and protection in the womb (notably Ps. 139:13–16; see also Ps. 119:73; Jb. 10:8; 31:15). Both Isaiah and Jeremiah refer to being called before birth (Is. 49:1, 5; Je. 1:5); indeed, Jeremiah's call came *before* he was 'formed in the womb', although this probably means nothing more than that God's pre-existent purpose was determinative for Jeremiah's destiny. In a similar way, our own choosing as Christians was 'before the creation of the world' (Eph. 1:4; 2 Tim. 1:9). Luke uses the same Greek word (*brephos*) to describe the unborn John the Baptist, the newborn Jesus, and the children brought to Jesus for blessing (Lk. 1:41, 44; 2:12, 16; 18:15). It is important, however, to note two points here.

First, all these passages involve an individual looking back to acknowledge God's care over him (or her) at all stages of life, and can apply only to the retrospect of a rational being. It is wholly illegitimate to argue that since God was involved with me from conception (never mind before conception), he is 'in' all conceptuses. Indeed, to do so is comparable to the mistake of arguing from his trustworthiness in arranging our destiny to the theological necessity for a geocentric Universe (Ps. 104:5; *etc.*).

Secondly, although a young foetus is biologically alive, there cannot be any firm certainty that it is also spiritually alive. Rather ironically, the strongest argument for according protection to foetuses from conception is based on our uncertainty of the spiritual status of the conceptus; it is worth noting that this argument is not a scriptural one.[2]

2. *Human life*. A previous generation tended to equate God's image as somehow dependent on rationality. We have already seen that this cannot be supported from the Bible; modern theologians agree that God's image is expressed relationally, not physically. In other words, our essential humanness is not located in our genes or any other specific physical component of us. We emphasize this in our evangelism when we proclaim that our response to God is a response of the will, though constrained by experience, family or genes (Jn. 1:13); logically, we should insist likewise that our created spiritual nature is independent of our genes. God's image in us is not inherited like other traits, or transmitted in the normal genetic fashion; it is implanted and transmitted according to God's own methods and purposes, which are not subject to the physical predictability and limitations of Mendelian segregation.

The apparent irrelevance of DNA to the image of God in us highlights a paradox in the emphases (on the one hand) of those who regard life as beginning at conception and (on the other) of those who maintain that some degree of physical organization is necessary before a developing foetus can be meaningfully regarded as human (for example, the attainment of a critical degree of neural complexity).[3] The 'life begins at conception' protagonists have to support their case by underlining the importance of the zygote formed by fusion of gametes, but at the same time they are forced to play down any physical attributes of the conceptus, lest they be taken as identifying the person-who-will-develop with particular traits.[4]

3. *Fall and atonement.* We have already discussed the effect of the fall on humans (p. 111), but it is pertinent to emphasize that although the fall had tangible effects, there is no general agreement on what they were. It is wrong, however, to claim that humans are in any sense fallen genetically. This becomes immediately apparent if we consider the implication of any genetical change produced by Adam's disobedience: it could in principle be reversed by genetical engineering techniques. This possibility is not dependent upon the limitations of available technology or on the morality of attempting such an interference. The point is that if the fall depended on a genetical change, it could be dealt with biochemically, with no need for Christ's redeeming work. It must necessarily be concluded that the fall was not a genetical event. This supports the idea that genes are, as it were, morally neutral.[5] It goes against (although it does not contradict) the possibility that fertilization (which is simply a fusion of paternally and maternally derived genes) is a morally crucial event.

4. *Marriage and children.* The Old Testament emphasis on the Jews as God's chosen race is a genetic concept, receiving backing from the command to increase in number and fill the earth (Gn. 1:28; 9:1), and the Levitical teaching on marriage and the importance of families. The importance of the Old Testament genetic line was abruptly and radically terminated, however, by the nature of the church founded by Christ, which is completely independent of any genetic links for its existence and spread (Jn. 1:12–13; Rom. 4:16; 1 Pet. 2:9; *etc.*). Indeed, Paul's exhortations to avoid arguments about genealogies (1 Tim. 1:4; Tit. 3:9) can be interpreted as warnings that genetic descent has been superseded by the new and living way of Christ.

It is proper to link this with marriage. This was ordained by God in the first place for companionship (Gn. 2:18, 24; Mt. 19:5). There is a strong theological tradition that the relational and procreational 'goods' of marriage can be separated only improperly (p. 163); Paul's emphasis on the 'one flesh' of marriage is part of this tradition (1 Cor. 6:16; Eph. 5:31). The teaching about 'one flesh' is not intrinsically about reproduction, however, but about the complementarity of male and female. The biblical usage of 'flesh' refers to much more than our physical body (Ps. 16:9; Pr. 14:30; Col. 2:11; *etc.*); Christ's view of adultery (Mt. 5:27–28) goes far beyond the legal definition of sexual intercourse. Scripture uses two words to describe the physical union of a man and a woman: *kollaō*, to join, glue or cement together, and *ginoskō*, to know. The two words describe the very deep emotional and spiritual relationship between a man and a woman, the physical vehicle and sign of which is sexual intercourse. On a purely pragmatic level, contraception separates sex from procreation, but most Protestants emphasize the creation interpretation of companionship, and take the 'one flesh' teaching as allowing them to practise contraception with no moral qualms.

This understanding of marriage clarifies the significance of third- (or fourth-) party gametes in a marriage, as happens when donor insemination (DI, previously known as AID – artificial insemination by donor) or ovum donation or both take place. AID has been ruled in a Scottish court not to be adultery (legally this was judged to involve the physical union of man and woman), even though to many it represents an illegitimate trespass into a marriage partnership (Ison 1983). But, and this is the significance in the present context, donor gametes are an unacceptable intrusion only if the genetic contribution has a major ethical importance. If genes are morally neutral, donor gametes can have no intrinsic disruptive influence on a marriage.

Dunstan has pointed out that condemnations of DI concentrate on 'the violation of nature and the breaking of the ordained nexus between sexual love, procreation and family life. In substance these arguments are consequential, and can be put to the test of experience and empirical proof' (Dunstan 1983). Such tests have only just began to be made, largely because the whole artificial-insemination procedure is surrounded, for understandable reasons, by secrecy.

Carl Whitehouse has written:

> Faced with the complex issue of what is involved in responsible parenthood, one turns to Scripture for guidance; one is then faced with the fact that, although there is a clear rule of lifelong marriage as the sole situation for sexual expression, there is less clarity on the right context for the rearing of children. At times one is amazed at the lack of condemnation of apparent aberrations. In Genesis we find a form of surrogate motherhood indulged in by the patriarchs; it may have been consistent with the social conventions of surrounding societies, but one is still surprised that it is not judged more harshly, for although Jacob's family had problems, the same can be said for strictly monogamous Isaac! Again, social parenting in Scripture was carried out on a more communal basis [than we are used to] and we do not consider Hannah an irresponsible parent because Samuel was sent to the temple as soon as he was weaned. In some ways Scripture, and this is borne out in

certain genealogies, holds the legal position of parent as of greater importance than the genetic . . . (Whitehouse 1983: 27).

In vitro *fertilization*

The possibility of 'test-tube' fertilization raises spectres of an engineered brave new world. But the ethical issues raised by *in vitro* fertilization (IVF) in the strict sense are no different from those in DI. Indeed, if the semen comes from the husband and the ovum from the wife, the procedure is genetically the same as artificial insemination by husband (AIH).

IVF, however, means that developing embryos are susceptible to study and manipulation outside the body. What are the limits to this? And more emotively, are investigations on early human embryos tampering with God's own prerogatives in relation to humankind?

The way forward would seem to be to acknowledge God's sovereignty over all life; to acknowledge that his image in us places a definite limitation on the extent to which we may manipulate human material; but at the same time to acknowledge that he expects us to use the skills we are given for proper ends, such as research on the causes of malformation and handicap. Only if we take an absolutist position, holding that any procedure must involve no risk whatsoever to an unconsenting child, can we decide differently with reason.

Conclusions

The crux of the Bible teaching about our biological nature is that God chooses and cares for people; he is not concerned with embryological or sociological niceties. An excellent example of this is the Lord's word to Jeremiah: 'Before I formed you in the womb I knew you' (Je. 1:5, RSV; *cf.* Ec. 11:5; Is. 44:2). This tells us about God's relationship to a person; it does not direct us to processes or to a scientific understanding of human biology, but neither did our Lord's admonition to Nicodemus about rebirth (Jn. 3:3–6).

Most Christian arguments about developments in reproductive medicine start from the premise that life is 'sacred'. This is acceptable if it means 'evocative of awe and commanding the utmost respect for the miracle that life is'. But if the word is used to mean absolute protection and inviolability, it becomes highly questionable. The traditional western ethical tradition has been to give a foetus increasing protection as it develops; to go beyond this because of a greater knowledge of embryological processes is to introduce a new rigour, which is not necessarily the most correct interpretation of our biological knowledge.

The point to emphasize is that good ethics are not likely to come from bad biology or naïve theology. The fact is that we are biological machines, but also God's creatures. The care of human life is a responsibility entrusted to us by God, and our attitude to the decisions that face us is an outworking of Christ's expectation for Christians, encapsulated in the teaching about salt and light in the Sermon on the Mount.

First, although the Bible says nothing about DNA, it says enough on related subjects to enable us to deduce a great deal about the apparent unimportance of genes in the divine economy. Our genome is an essential part of who we are, but it is not all-encompassing: in God's sight we are called to respond with all that we are to his transforming grace: and it is grace which is the crucial factor. This does not mean that our genetical composition is to be ignored or neglected (many of our God-given talents are latent in our genes), but rather that we have to recognize that the divine factor or image does not reside therein.

Secondly, the change from what may be called the classical assessment of the early foetus resulted from a better understanding of embryology on the one hand and concern about easier abortion on the other (as well as alterations in papal dogma); in other words, from scientific or technical developments rather than biblical interpretation. The assertion that 'life begins at conception' has a secular derivation, and we need to note that it produces problems at the ethical interface, for the following reasons:

1. It tends to imply a divine determinism in development, so that inherited traits are regarded as fixed and God-given (or, alternatively and paradoxically, to be a result of the fall). This extends (and distorts) the traditional interpretation of 'natural law'.

2. It is faced with the difficulty of interpreting the status and fate of the 70–80% of foetuses which fail to develop normally, including hydatiform moles, which are conceptuses even though they are formed from the fusion of two sperm nuclei with an ovum cytoplasm, and never produce a viable embryo.

3. It has problems in assessing the spiritual status of identical twins, which are formed by the division of a single embryo at any time up to ten days or so after fertilization (and likewise the rare embryos which result from the fusion of two zygotes or early blastulae).

4. It fails to take into account the commonly accepted consensus about the end of life, that is that death is the result of brain death rather than any measure of heart or kidney or any other physiological function. There is a good case for the analogy of regarding individual life as beginning when a certain minimum level of brain activity is established.

Thirdly, there is a danger that the elaboration of ever more sophisticated techniques of manipulating the human body leads us to treat it as merely a machine, rather than as a vehicle for God's Spirit. This is particularly acute for those who accept some degree of dualism between body and spirit. In recognizing and acknowledging that we are 'fearfully and wonderfully made', we need to remember also the complementary truth that our Maker is also the creator and redeemer of the whole world, and that 'all things hold together in him'.

Finally, both biologically and theologically we are unique. This is not a matter for regret or repentance, but for seeking and working out God's purpose for us as individuals.

Chapter 10

Brain, mind and behaviour

At the end of the twentieth century, some of the most exciting scientific developments are in neuroscience and related disciplines. Today's discoveries are properly understood only against the background of earlier views about how the mind relates to the body and, in particular, to the brain. The accumulating evidence points to the tightening of the mind–brain link. What does that imply for our conviction that we are free to make choices? Whatever has become of the soul? What of that most mysterious of personal qualities, our consciousness? Is it likely to be explained in biological terms and, if so, what might that imply about human uniqueness? Leaders in research hold differing views on these issues. What are we to make of it all and what might it imply for some of our Christian beliefs about human nature?

Is sexual orientation fixed by our genes? If not, what degrees of freedom are there for the expression of our sexual behaviour and how responsible are we for it? Are our spiritual experiences modulated by our biological make-up? What, for example, is happening when a devout Christian, in the terminal stages of Alzheimer's disease, blasphemes and denies his or her faith?

The last three decades of the twentieth century have seen some of the most exciting developments ever witnessed in neuroscience and neuropsychology, and all apparently tighten the link between brain, mind and behaviour. We shall briefly review the main features of these converging lines of scientific evidence and ask what they imply. What sort of an emerging picture of humankind do they suggest, and how do we relate this to the salient features of a biblical view?

Converging lines of scientific evidence

For the great anatomists Gall (1758–1828) and Spurzheim (1776–1832) it was scientifically respectable to attribute differences in abilities and personality to variations in the size of the convolutions of the brain. 'Phrenology' was a conventional scientific discipline. It was not long, however, before it fell into disrepute, and for good reasons. Experiments with animals seeking to localize specific mental functions to particular parts of the brain were unsuccessful. The dramatic case of Phineas Gage began a new approach.

Phineas Gage was a twenty-five-year-old foreman working in summer 1888 for the Rutland and Burlington Railroad Company in New England, and was described by his employers as 'most efficient and capable'. As he prepared an explosion to remove rock obstructing the path of the railroad, someone called to him from behind and he briefly looked away. It was only for an instant, but when he turned back and began tamping the explosive with his iron bar, without realizing that his assistant had not poured in sand beforehand, a deafening explosion took place. The bar, weighing 6 kg, over 1 m long, and tapered to a point 6 mm in diameter, entered Gage's left cheek, pierced the skull, crossed the front of his brain and emerged through the top of his head. The rod landed more than 20 m away, covered in blood and brains. Phineas Gage was stunned but amazingly still conscious. He went on to make a remarkable recovery, but his personality had undergone a dramatic change. His likes and dislikes, his aspirations, his ethics and morals were all different.

Earlier studies of neurological damage had begun to show that the brain may be the foundation for language, for perception and for motor control, but Phineas Gage's accident suggested that there are also systems in the human brain which affect the personal and social dimensions of normal life. He lost previously acquired social conventions and ethical rules as a result of selective brain damage, even though his basic intellect and language remained essentially normal. This showed the existence of a link between the intactness of several aspects of mind: attention, perception, memory and language as well as personality and character. Debate raged around whether and to what extent the damage caused to Gage's brain was localized, and the degree to which such localization was linked to his character change. A leading phrenologist, Nelson Sizer, claimed that the iron bar had passed 'in the neighbourhood of Benevolence and the front part of Veneration'; in other words, that there were brain centres which were the substrate for proper behaviour, including kindness and respect for others. In the case of Gage, these had been selectively injured. With hindsight it is clear that Gage's personality change was brought about by a circumscribed brain lesion in a specific site, but the important lesson was that to observe social convention, to behave ethically, and to make proper decisions requires the integrity of specific brain systems. Antonio Damasio asks, 'Is it fair to say that his soul was diminished, or that he had lost his soul?'

In reflecting on Gage, Damasio writes:

> Gage's story hinted at an amazing fact: Somehow, there were systems in the human brain dedicated more to reasoning than to anything else, and in particular to the social and personal dimensions of reasoning . . . Unwittingly, Gage's example indicated that something in the brain was concerned specifically with *unique human properties*, among them the ability to anticipate the future and plan accordingly within a complex social environment, *the sense of responsibility* toward self and others, and the ability to orchestrate one's survival deliberately, *at the command of one's free will* (Damasio 1994: 10).

Damasio has identified modern Phineas Gages who can easily be missed and mis-diagnosed. Based on his clinical experience, Damasio comments: 'The distinction between diseases of "brain" and "mind", between "neurological" problems and "psychological" or "psychiatric" ones, is an unfortunate cultural inheritance *that permeates society* and medicine. It reflects a basic ignorance of the relation between brain and mind' (1994: 40).

Determinism

At this point, we need to enquire about the implications of linking brain and behaviour. In a rough and ready sense, we all take it for granted that human behaviour is predictable. If you simply reflect back on the last twenty-four hours, you will find that what you did and said was largely predictable. In this sense at least, psychological determinism is not some kind of special demonic prejudice held by atheist materialist psychologists, but is an implicit working assumption that most of us hold most of the time. In practice, determinism amounts to little more than acknowledging that human behaviour exhibits regularities (laws), that these regularities seem to be susceptible to rational causal explanations (theories), and that for this reason it is sensible to set up our research programmes in order to push our understanding of these laws and theories to the limits. If a particular piece of behaviour seems capricious, our natural reaction is to subject it to more careful study, hoping to find the hidden laws which we expect *really* to apply.

Some who criticize contemporary psychology do so on the grounds that it makes a *metaphysical* assumption of determinism. As we have seen in discussing reductionism (p. 76), however, there is an important difference between what can be called *methodological* determinism and an inflexible and absolute ontological presupposition about the way the world is. Almost all scientific psychologists are methodological determinists. If they did not aim to identify regularities in human behaviour that would enable them to explain and hence predict it, there would be little point in their whole enterprise. Why undertake an experiment on how schoolchildren learn mathematics, or why particular drugs affect the efficiency of industrial workers, or what are the relative values of different types of psychotherapy, if their cause is intrinsically capricious? If that were so, then the 'laws' being looked for in each experiment would not really be laws at all. The results of any experiment would not apply to further instances of the same kind of situations.

If the tentative application of a methodological determinism turns out to be reasonably successful, then the next step is to assume that we can predict particular behaviours, and hence that exceptions are probably due to a lack of complete information. This stage has sometimes been called *empirical* determinism.

If we confined ourselves to *methodological* and *empirical* determinism, we would be in the position that was popular in the 1940s, claiming that all human acts are mere extensions of the laws of physics and chemistry, and hence that we have a limited ability to influence behaviour. Such a conclusion seems to some to remain inevitable from subsequent decades of study on the neural basis of mental life, which has pointed to an ever-tightening link between mind and brain. Indeed, it is not unusual to be told that we are no more than complexes of processes which can in principle be formulated in physical and chemical terms. In other words, descriptive and causal statements made in psychological language can in principle be translated without loss into statements in physiological language. This is effectively an assumption of what may be called *metaphysical* determinism. The serious implications of this position have stimulated psychologists who are also Christians to re-examine the determinism of human behaviour, particularly the question of whether we do in fact enjoy any real freedom of choice.

The mind–brain link and Christian faith

One way of approaching the issue of determinism and the link between mind and brain is by reminding ourselves of the intimate connection between disordered brain chemistry and change in behaviour, such as the so-called dopamine hypothesis concerning some forms of schizophrenia. This is important because some people who have manifested apparent signs of God's work in their lives may turn out to have been suffering from diseases such as epilepsy or schizophrenia. For Christians who believe they have freely responded to God's loving invitation to put their trust in him and to live their lives according to his teaching, any suggestion that they do *not* have free will must be resisted. For example, there are those who suffer spiritual distress because what begins as neural degeneration in the brain leads on to psychological disordering of the mind, and this in turn may have profound spiritual consequences. Alzheimer's disease is an example. Like the rest of the body, the brain is affected by the passing of the years. We all experience the occasional lapse of memory, but in some this turns into senile dementia. It may affect as many as 5% of the over-sixty-fives, and up to 30% of the over-eighties and nineties. Dementia is a consequence of a variety of different diseases, but Alzheimer's is by far the most common.

Many of the symptoms of Alzheimer's dementia are caused by damage to brain cells. Neurofibrillary tangles develop in the brain; granulovascular deterioration becomes evident in the brain cells. In addition, there are neurochemical changes, most often a deficiency of the neurotransmitter acetylcholine. One result is that previously patterned circuits of nerve signals become scrambled and the brain can no longer sustain its normal psychological functions.

These changes in brain structure have predictable psychological consequences. The major symptom is loss of memory, leading to confused language, thinking and recognition. Other upsets are in perception, in feelings and in motor activity. At the last comes what is traditionally labelled dementia. At this stage, sufferers are apt to wander and are incapable of basic self-care; they need to be dressed, washed, fed and helped with bowel functions. They continually relive memories from the far distant past, which are much more accessible to them than memories of recent events; they often become paranoid, making false accusations about assaults on their person. The result is psychotic-like delusions and hallucinations. Finally may come seizures and ultimately coma and death.

In some Alzheimer's patients, those who knew them well witness apparent changes in personality. These can be most distressing. For example, a theologian, previously a devout believer, became profane in his language and said he did not believe in God. When the physical substrate disintegrates, the mental processes go all awry.

How can we best think about the subtle and intimate relationship between mind and brain? Clearly, both matter, and we need knowledge at several different levels if we are to begin to understand the complex processes occurring as someone undertakes the simplest of cognitive tasks. Explaining what is happening at one level is not the same as explaining away the phenomenon under investigation (pp. 76–77). Damasio is correct to ask:

> Does this mean that love, generosity, kindness, compassion, honesty, and other commendable human characteristics are *nothing but* the result of

conscious, but selfish, survival-oriented neurobiological regulation? Does this deny the possibility of altruism and *negate free will?* Does this mean that there is no true love, no sincere friendship, no genuine compassion? That is definitely *not* the case . . . Realizing that there are biological mechanisms behind the most sublime human behaviour does not imply a simplistic reduction to the nuts and bolts of neurobiology (1994: 125–126; our italics in the first two cases).

Mind–brain: dual aspects of a single reality

There is more than one way of interpreting the evidence on the tightening of the mind–brain link. Until relatively recently, the gap between single-cell studies by neuroscientists and cognitive studies by psychologists seemed so large as to be virtually unbridgeable. There are some neuroscientists who believe that the only way to make progress is to concentrate exclusively on the study of the most basic processes of neural functioning – what is sometimes called the 'bottom up' approach; other scientists are equally convinced that the only way forward is to concentrate on gaining a deeper understanding of mental processes at the level of investigation of cognitive psychology – a commitment to a 'top down' approach.

The need to recognize and do justice to what is happening at the higher levels of analysis led Nobel Laureate brain scientist and psychologist Roger Sperry to declare that a major revolution in our thinking about mind and brain has taken place, identifying 'a move away from the mechanistic, deterministic and reductionistic doctrines of the pre-1965 science to the more humanistic interpretations of the 1970s' (Sperry 1988: 11). As we shall see later, while Sperry's own formulation of the mind–brain relationship is not without its problems, the fact remains that after working at the leading edge of the discipline he firmly believed that it is simplistic to try to reduce humans to 'nothing but' physico-chemical machines.

A range of options faces us as we consider how best to characterize the complex interactions between mind and brain. It is not unusual to make a set of metaphysical assumptions, not always openly declared. Those with a religious commitment often espouse some form of *dualism*. On this view, mental events act causally on the brain, at times preceding the corresponding brain activity.

Dualism assumes that the body is made of physically determinate stuff, but maintains that mind enjoys a certain autonomy of its own. Such a belief is often, though not exclusively, held by those with interests in subjects such as parapsychology, extrasensory perception, psychical research, and so on. It is a defensible view. In 1950 Peter Laslett edited a book on *The Physical Basis of Mind*, in which a number of distinguished biological scientists reached something approaching a consensus that the neural activity of the brain somehow *interacts* with the private world of the mind. The question remained, however, to what extent it was necessary to hold a dualist view in order to safeguard individual freedom.

Another form of dualism has been labelled 'dualism of descriptive categories'. On this view, freedom and determinism are concepts belonging to two different language systems. Its advocates argue that both are necessary to do justice to our present scientific knowledge and to the experience of human freedom. For example, the theoretical physicist Nils Bohr speaks of freedom and determinism as complementary descriptions, drawing attention to the analogy of wave–particle models in physics. Certainly this view avoids some of the difficulties of the 'dualism of substance' approach. Others simply point out that, while determinism is a useful postulate *within science*, that does not mean that it is a universal rule about the world. It is one thing, they argue, to employ it as a useful rule of procedure for scientific enquiry, but quite another to assert that such a rule expresses an intrinsic property of the created order. Among psychologists, Professor Carl Rogers takes this view. He writes that 'responsible personal choice is the core experience of psychotherapy, and exists prior to any scientific endeavour . . . To deny the experience of responsible choice is as restricted a view as to deny the possibility of behavioural science' (1980).

Yet another variety of dualism is advocated by some who see a solution to the problem within science itself. Their appeal is to the Heisenberg principle of indeterminacy in physics. For example, the physiologist Sir John Eccles believes that some form of mind–brain liaison occurs in the cortex (that is, in the large convoluted surface of the brain's cerebral hemispheres), and influences neural activity without violating physical laws because the energy involved in such influence is within the limits of the Heisenberg uncertainty principle.

Eccles's view has not found wide acceptance. As we have seen, first, the indeterminacy allowed by the Heisenberg principle becomes increasingly negligible with bigger objects (p. 92). For electrons its effect is far from negligible, but by the time we get to the neuron, which is a million million times heavier than an electron, it is virtually trivial. Secondly, the brain seems to be organized on a teamwork basis, so that unpredictable behaviour of one brain cell would make no significant difference to the overall functioning of the brain. Thirdly, the random fluctuations in the brain attributable to Heisenberg's principle are extremely small compared with other fluctuations known to us, such as those due to thermodynamic changes, to random fluctuations in the blood supply, and so on. In fact, we have to conclude that such unpredictable disturbances could as easily be used to excuse subjects from responsibility as to credit them with responsibility for their choices.

One non-dualist interpretation simply identifies mind with brain. Such an approach is linked to materialist presuppositions, typified by behaviourists of an earlier generation such as B. F. Skinner. They argued that mental events are a product of physical events but have no causal efficacy: such views are labelled *epiphenomenalist*. It is characteristic of all varieties of *materialist* views that they give ontological priority to matter – in this instance, brain rather than mind. There is one interesting twist to the views of the thoroughgoing *identity* hypothesis sometimes overlooked; since they do not deny the *reality* of the physical substrate, and since this is by their definition identical with the *mental* events, therefore the mental events are also *real*.

Yet others believe that the best way of thinking about mind–brain links is *psychophysiological parallelism*. On this view, there are two parallel streams of events. Physical events cause further physical events, and mental events further mental events. The physical and mental events are tightly coupled in time.

The view we take was labelled by Donald MacKay as *comprehensive realism*. Here mental activity and correlated brain activity are seen as inner and outer aspects of the same complex set of events which together constitute conscious human agency. The two stories that may be written about such a complex set of events, the mental story and the brain story, are said to demonstrate logical complementarity (see chapter 4 for discussion of this). The irreducible duality of human nature is, on this view, seen as duality of aspects rather than duality of substance.

Thus the tightening of the link between mind and brain need not reduce the importance of the mind or of mental activity in general, nor does it mean that the mind is a mere epiphenomenon of the physical activity of the brain. The mind determines brain activity and behaviour. But in a complementary fashion mental activity and behaviour depend upon the physically determinate operations of the brain, which is a physico-chemical system. When that system goes wrong, there are changes in its capabilities for running the things we describe as the mind or mental activity. And, likewise, if the mind or the mental activity results in behaviour of particular kinds, this may result in temporary or chronic changes in the physico-chemical make-up and activity of the brain, its physical substrate.

We therefore regard mental activity as being embodied in brain activity rather than as identical with brain activity. To go beyond this to what is called a monist identity view confuses categories which belong to two different logical levels. There is nothing within brain science or psychology which offers any justification for asserting that a monist identity view fits better with the evidence than the view we have outlined here. Roger Sperry put it well: 'The laws of biophysics and biochemistry are not adequate to account for the cognitive sequencing of a train of thought' (1988). Professor J. Z. Young has argued similarly: 'the brain contains programmes, which operate as a *person* makes selection among the repertoire of possible thoughts and actions' (1987: 19). Note it is the *person* who makes the selection, *not* the brain. We have continually to be aware of the danger of confusing talk about brains or machines with talk about persons. It is conscious cognitive agents who think; it is people, *not* brains, who make choices. Damasio echoes this view: 'To understand in a satisfactory manner the brain that fabricates human mind and human behaviour, it is necessary to take into account its social and cultural context. And that makes the endeavour truly daunting' (1994: 260).

John Polkinghorne refers to 'the perpetual puzzle of the connection of mind and brain' (1986: 92). He points out that if you are a thoroughgoing reductionist, the answer is easy: 'mind is the epiphenomenon of brain, a mere symptom of its physical activity . . . but the reductionist programme in the end subverts itself. Ultimately it is suicidal' (1986: 92). It destroys rationality, and thought is identified with electro-chemical neural events. Such events cannot confront one another in rational discourse; they are neither right nor wrong; they just happen. Polkinghorne writes: 'If our mental life is nothing but the humming activity of an immensely complexly connected computer-like brain, who is to say whether the programme running on the intricate machine is correct or not? . . . The very assertions of the reductionist himself are nothing but blips in the neural network of his brain. The world of rational thought discourse dissolves into the absurd chatter of firing synapses. Quite frankly, that cannot be right and none of us believes it to be so' (1986: 93). He echoes J. B. S. Haldane, who, many years earlier, wrote: 'If my mental processes are determined wholly by the motions of the atoms in my brain, I have no reason to suppose that my beliefs are true . . . and hence I have no reason for supposing my brain to be composed of atoms' (1927: 209).

One cannot emphasize too strongly the dangers of sliding into a form of thinking about human beings which relegates the conscious cognitive agent to second place, or, as Lipowski put it, of adopting 'the reductionistic gospel'. The point is simply that *all* our knowledge of brains and minds and computers comes *only* through our experience and activity as conscious agents. It is only in and through that experience that we gain scientific or any other kind of knowledge; that means that the *conscious agent* has what philosophers call ontological priority. As Donald MacKay put it, 'Nothing could be more fraudulent than the pretence that science requires or justifies a materialist ontology in which ultimate reality goes to what can be weighed and measured, and human consciousness is reduced to a mere epiphenomenon.' No amount of analogies, however, of this or any other kind, ought to reduce our awe as we reflect upon our own experience as embodied conscious agents. The sense of mystery, for us at least, remains untouched by any amount of brain science.

What constitutes freedom to choose?

At this point we need to digress and ask what we mean by free will and choice. We need hardly say that this is a topic with a very long history. Relevant to our present discussions is the reminder that there have been two distinct definitions of freedom, the liberty of spontaneity and the liberty of indifference. To understand the difference between the two, consider a person faced with a choice at breakfast of porridge or stewed fruit. Imagine she chooses the porridge. Now if we were able to set up again exactly the same circumstances as were in force at the moment the choice was made, including a specification of the state of the whole Universe, not forgetting the person's brain, then two possibilities exist: (1) the liberty of spontaneity, according to which the person would always choose porridge, since choosing porridge is what she wanted to do; and (2) the liberty of indifference, according to which the person would have the ability to take either porridge or stewed fruit. The first, the liberty of spontaneity, is often referred to as a compatibilist view of freedom. This is because it is compatible with determinism. The liberty of indifference is referred to as a libertarian view of freedom.

The liberty of indifference

Christian scientists such as John Polkinghorne and Arthur Peacocke seem to construe freedom in terms of the liberty of indifference. John Polkinghorne believes that anything less than this would relegate us to being automata. If, as he assumes, we accept a non-dualist view of mind, then independence from all physical circumstance seems to imply an independence from the previous state of the mind. The problem then becomes: in what sense can the choice be said to be caused by the mind? In effect, it looks as if this definition of freedom gives us too great an independence in that it produces an independence from our own selves; we are not the source but only the scene of the choice. This position seems to revert to dualism.

Polkinghorne has argued that the liberty of indifference is necessary if we are to understand our rationality, for 'if the brain is a machine, what validates the programme running on it?' (1994a: 12). He argues that human rational judgment has to enjoy an autonomous validity in order to discern the truth, and that this would be negated if it were the by-product of mere physical necessity. Notwithstanding, we need to remember that 'mere physical necessity' is the unbelievably complex system of the brain. It looks as though the autonomy of objectivity which Polkinghorne wishes to defend is more than the autonomy from past circumstance that the liberty of indifference provides. What, we must ask, would validate the truth-discerning capacities of this freedom? It looks as though the explanation of our rationality is difficult whatever view of freedom we take. It is indeed, as Polkinghorne has pointed out, hard to conceive how evolutionary necessity would ensure the subtlety and fruitfulness of human reason' (1994a: 12).

The liberty of spontaneity and logical indeterminacy

The late Professor Donald MacKay gave the whole debate on free will a new twist by defining freedom in a new way, one that is consistent with the liberty of spontaneity.

In presenting the view, MacKay asks us to imagine the following scenario (1991: 194–204). Consider a person, call him *A*, whose behaviour we are studying and whose next choice we seek to predict. We can think of a super-scientist who is able to look into *A*'s brain, see its complete state and, on the basis of this, so it is claimed, be able to predict the future brain state of *A*. But, asks MacKay, would *A* himself be correct to believe the prediction being made by the super-scientist if it were offered to him? His answer is no. MacKay argues that if the super-scientist's predictions were presented to *A*, then it would in fact alter *A*'s brain state, and so the prediction would immediately be invalidated. Since by definition no change can take place in what *A* believes without a correlated change in his brain state, it follows that no completely detailed specification of the present or immediately future state of his brain could be correct whether or not *A* believed it. If it were correct before *A* believed it, then it must be incorrect in some detail after *A* comes to believe it; conversely, if the specification were adjusted so that it would become correct if and when the brain state were changed by *A*'s believing it, then it would not be correct unless *A* believed it.

MacKay then goes on to a further situation which supposes that our super-scientist could in fact formulate a new prediction which would take into account the effect of offering it to *A* and would be correct then only if *A* believed it. The question then is: would such a prediction command *A*'s unconditional assent? MacKay answers no. He claims that *A* would obviously be entitled to believe it. *A* would, however, be equally entitled to disbelieve it, as a prediction is true of and only if *A* believes it. Putting this in a more formal way, we can therefore say there does not exist a complete description of *A*'s future which would have an unconditional claim to *A*'s assent, in the sense that *A* would be correct to believe it and mistaken to disbelieve it, if only *A* knew it. MacKay argues that in this sense *A*'s future is logically indeterminate, and that from *A*'s point of view the future is open and up to *A* to determine. For MacKay, this is the essence of freedom, as he asserts that it is not brains but persons who choose. It is at the level of our conscious experience that there is an indeterminacy, irrespective of any indeterminacy at the level of the brain. MacKay remained uneasy with any definition of freedom which did not explicitly distinguish between these two levels, and he suggested that it is a mistake to attempt to secure our freedom by trying to exploit physically indeterminate processes in the brain, along the lines of Eccles's arguments (p. 177).

In presenting his argument, MacKay was careful to opt for the worst-case scenario. As he put it in his Gifford Lectures (*Behind the Eye*), 'We . . . are going to be asking, what if – and it is a strongly underlined if – the whole of this system as summarized here were a determinate system in the physical sense? That is to say, if every physical event had its adequate determinants in other and earlier physical events, would it follow that we have no choice, everything is inevitable, and we couldn't have done otherwise?' (1991: 193). MacKay is here arguing that if we can demonstrate freedom in such a restricted world, then, *a fortiori*, it can certainly be established in the real world. He makes it clear elsewhere that in fact he believes it is highly unlikely that the world is totally physically determinate. The second point, which is frequently misunderstood, is that the argument is not concerned with what *A* would actually do when he was offered the prediction, but rather making a logical point about what it would be correct for *A* to believe. MacKay is also emphasizing that his logical-indeterminacy argument applies to statements that involve what he calls the cognitive mechanism, that is to say the part of the brain embodying a person's beliefs. Indeed, he suggests that we should be held responsible only for those actions which are physically dependent on our cognitive mechanism. And MacKay also believes that the way he defines freedom is fully compatible with theological determinism, that is, in a universe where God determines all that is and, therefore, divine sovereignty and human freedom are reconciled.

Donald MacKay's argument has been discussed and critiqued by so many logicians and philosophers as well as scientists, that there is no doubt that the logical point he wished to make is firmly established. For MacKay's critics, the problem centres on whether it is proper to equate the notion of freedom with logical indeterminacy in the way that he does.

It is important to stress that MacKay had no desire to appeal to unpredictability as a means of retaining the kind of freedom he believed is required. Indeed, he commented, 'In so far as quantum events disrupted the normal cause and effect relationships between the physical correlates of my rational deliberation, such events might be held to diminish rather than enhance my responsibility for the outcome' (1974: 94). To attempt to retain the idea of freedom by an appeal to physically indeterminate processes produces the further question of how, in that case, the choice that is made can be attributable to a personal agent. John Polkinghorne has tackled this problem differently, by referring to the inherent openness of complex dynamical systems. For him the future would be contained within an envelope of possibilities so that the actual pathway followed could be selected by input of confirmation by the mind. This sounds like the introduction of a new form of dualism by the back door. It is a kind of 'top down' operating causation that for Polkinghorne could be the locus of human freedom. He has elsewhere argued strongly for the kind of dual aspect monism which we adopted elsewhere in this book. On the face of it there then seems to be some conflict between his acceptance of a dual aspect monism and his attempt to exploit the openness of complex dynamic systems in the way that he does.

A problem with Polkinghorne's view is that it is difficult 'to attribute the information input to an act of the mind, it needs to be preceded by a mental decision on the desired outcome. But if the mind is embodied, this decision would already have a physical correlate and so the information input cannot be the point of choice. Freedom must lie elsewhere. Alternatively, if the information input is not seen to be the result of a previous state of mind, then we get back to the original problem of the liberty of indifference — what causes this information input, chance? Its attempts to explain the freedom of a non-dualistic mind through the openness of physical processes are likely to meet this problem' (Doye *et al*. 1995: 127). There are difficulties also with the views of MacKay and Peacocke: 'Put starkly, the following questions remain: is the liberty of spontaneity really free, and is the liberty of indifference really a result of our willing? Neither view appears to be philosophically or empirically necessitated and consequently, it needs to be opened to review in the light of its implications in other areas, particularly theology. However, we note the general point that the liberty of spontaneity can be more easily reconciled with God's sovereignty, than the liberty of indifference' (Doye *et al*. 1995: 127–128).

As a not unimportant footnote to this discussion, we would wish to suggest that the term 'sovereignty' requires careful re-examination. 'Sovereignty' is a term drawn from the royal metaphor (God portrayed as king), and this focuses our attention primarily not on power but on authority. The term 'sovereignty' therefore should be seen as a matter of having ultimate authority rather than of exercising ultimate control. Because God is sovereign, we are responsible to him and answerable to him. When sovereignty is pictured, not in terms of authority to which we are answerable, but rather in terms of manipulative or coercive control, as it so often is, the essence of the royal metaphor has been severely distorted.

Consciousness restored

Consciousness, a topic of prime interest to psychologists a century ago, almost disappeared for many years as an issue to research into or to write about. Today, scientists, almost without exception, acknowledge its mysterious nature. Thus Francis Crick wrote that his book *The Astonishing Hypothesis* (1994: xi) was 'about the mystery of consciousness'; Sir Roger

184   Science, life and Christian belief

Penrose marvels over 'how a material object (a brain) can actually evoke consciousness' (1989: 523); Euan Squires, author of *Conscious Mind in the Physical World*, puzzles 'where, within physics, does mind and consciousness appear? . . . Can physical things be conscious?' (1990); Sir John Eccles writes of the 'tremendous intellectual tasks [that face us] in our efforts to understand baffling problems that lie right at the centre of my being' (1966: 327).

Attempts by physical as well as biological scientists to understand consciousness illustrate how scientific data do not come neatly labelled with interpretation and meaning. As regards the biologists, it is worth noting the differing views of four Nobel Laureates in medicine and physiology. First, Francis Crick, a self-confessed materialist reductionist, happily asserts: 'You are *nothing but* a pack of neurones . . . *no more than* the behaviour of a vast assembly of nerve cells and their associated molecules' (1994), and 'consciousness depends crucially on thalamic connections with the cortex. It exists only if certain cortical areas have reverberatory circuits (involving cortical layers 4 and 6) that project strongly enough to produce significant reverberations'. He goes on, however, to say, 'I hope nobody will call this the Crick Theory of Consciousness . . . If anyone else produced it, I would unhesitatingly condemn it as a house of cards. Touch it, and it collapses . . . because it has been carpentered together, with not enough crucial experimental evidence to support its various parts. Its only virtue is that it may prod scientists and philosophers to think about these problems in neural terms, and so accelerate the experimental attack on consciousness' (1994: 252).

By contrast, consciousness for Sir John Eccles 'is dependent on the existence of a sufficient number of such critically poised neurones, and, consequently, only in such conditions are willing and perceiving possible. However, it is not necessary for the whole cortex to be in this special dynamic state' (1989: 327). And thus 'we can face up anew to the extraordinary problems inherent in a strong dualism. *Interaction of brain and conscious mind, brain receiving from conscious mind in a willed action and in turn transmitting to mind in a conscious experience* . . . let us be quite clear that for each of us the primary reality is our consciousness – everything else is derivative and has a second order reality. We have tremendous intellectual tasks in our efforts to understand baffling problems that lie right at the centre of our being' (1989: 327). Eccles's conclusion is that 'So far science cannot explain consciousness so this is where God is to be seen at work' (1966: 327). We dissent strongly from this; it is merely to go down the God-of-the-gaps route (p. 79).

The third Nobel Laureate is Gerald Edelman. What he finds so daunting about consciousness is that 'it does not seem to be a matter of behaviour. It just *is* – winking on with the light, multiple and simultaneous in its modes and objects, ineluctably ours. It is a process and one that is hard to score . . . Why is there a mystique about consciousness? A reasonable answer seems to be that each consciousness depends on its unique history and embodiment. And given that a human conscious self is constructed, somewhat paradoxically, by social interactions, yet has been selected for during evolution to realize mainly the aims and satisfactions of each biological individual, it is perhaps no surprise that as individuals we want an explanation that science cannot give' (1992: 111).

Finally, for Roger Sperry, 'consciousness is conceived to be a dynamic emergent property of brain activity, neither identical with, nor reducible to, the neural events of which it is mainly composed' (1988: 312), and again, 'consciousness exerts potent causal effects in the interplay of cerebral operations' (1988: 312).

Bringing together these four views, we note the recurrent mention of the 'mysterious' nature of consciousness. All four agree that it will not do to pretend that consciousness is merely something secondary to our other activities as human beings and as scientists; our endeavours and achievements in any walk of life are dependent ultimately upon this remarkable phenomenon of individual consciousness. If ever the ontological priority of mind needed to be emphasized, it is forcefully achieved in the varied approaches of these four scientists, all sharing a conviction that consciousness is a proper topic for scientific investigation.

Clearly, there will be no quick and easy answers available from the scientific study of consciousness. Even so, there are enough data to warn against accepting either dualism or eliminative monism, which have been and continue to be widely canvassed by some. Jean Delacour wrote:

> The sterility of dualism has been stressed repeatedly and it is now a classical introduction to a text on consciousness with Descartes playing the role of a 'straw man'. The insufficiency of eliminative monism is less obvious, since it can be easily masked by scientific arguments; moreover, this philosophical position fits well with reductionism, which until recently was dominant in neurosciences. However, consciousness as a fact of experience withstands any form of reductionism (1995: 1061).

In short, neither dualism of substance nor reductionism will do.

Despite Crick's attempt to set up a conflict between the scientific approach and what he takes as the religious view of human nature (which, he believes, argues for a 'hypothetical immortal soul' and involves having 'souls, in the literal and not merely metaphorical sense'; 1994: 260–261), we believe there is no necessary conflict between the biblical view of human nature (as outlined in chapter 8) and the emerging picture from scientific research. Both point to one set of physical events which require analysis from two (at least) distinct perspectives in order to begin to do justice to what is being discovered scientifically and to our primary experience as conscious cognitive agents.

Two conclusions emerge from this review of consciousness. First, conscious experience is no longer regarded as being too subjective, too private and therefore too far removed from the relative certainties of physics, chemistry or cell biology to warrant serious study. Conscious experiences are certainly private and are frequently subjectively tinted, *but they are the bedrock of all experience*, and it is the only place from which any of us can start. We shall certainly want to ask the views of our colleagues and request them to repeat our observations and see whether they confirm our findings, while remembering that *their* reports are about *their* very private experiences. The same happens whether one looks down a microscope, observes the colour of a liquid, or takes a reading from a scientific instrument. It is here and here only that science can begin. Of course, there are times when, for whatever reason, we may suffer from hallucinations, but *my* existence, and *my* experience as a conscious agent, remain for me the most certain features of my life.

Secondly, there is, currently, intensive scientific activity dedicated to developing a greater understanding of consciousness, what it is and what is its neural substrate. The data, however, do not come nicely labelled with their interpretation. Equally competent, equally distinguished scientists from different disciplines and with different presuppositions look at the same set of data and arrive at different conclusions. For our part, we take the view that it is unnecessary and unwise (as well as unbiblical), to adopt a *dualist* assumption of the relation of mind and brain. A more parsimonious and scientifically probable view is that, to do justice to the complexity of the one set of neurochemical brain events occurring during conscious experience, we must give ontological priority to mental life. As Christians, we believe that we can be peaceably open-minded about questions such as whether or not consciousness occurs in other animals beside ourselves. The work of primatologists certainly points to the presence of mind-like conscious reflective and deceptive behaviour in the higher non-human primates. There are no problems here for the biblically informed Christian who recognizes that we share many properties with animals, particularly the apes.

Certainly the scientist who is a Christian can be enthusiastically committed to, and involved in, work at the cutting edge of neuroscience and neuropsychology aimed at elucidating the mysterious nature of our ubiquitous experience of consciousness. In so doing we shall, as Crick points out (perhaps surprisingly), be filling out a little more of just how 'fearfully and wonderfully made' we really are. Moreover, we shall continue to be, as Edelman reminded us, keen that 'Everything in scientific enquiry should be exposed to remorseless criticism' (1992: 97).

Brain and behaviour and genetics

There are those who believe that complex social phenomena ranging from personal happiness to racial discrimination are in principle reducible to basic brain chemistry and/or disordered genes. For them, drugs like Prozac are the answer to unhappiness. Such views arise when advances in neuroscience are linked to a misunderstanding of the genetic basis of behaviour. From such a link, these reductionists smuggle their personal ideologies into explaining away aspects of behaviour which they wish to condone or to encourage.

For example, one view of homosexual behaviour is that a person is homosexual because of 'gay genes' which produce a 'gay brain', and that this licenses homosexual behaviour. Going down that road, it is easy to argue that for similar reasons street violence arises because certain people have reduced serotonin levels in their brains, resulting from violent or criminal genes, or that people get drunk because they have genes for alcoholism. Such a distortion of the meaning of the biological evidence occurs when some of the excellent studies of the neural and genetic bases of behaviour, carried out using experimental animals such as rats and cats, are extrapolated lock, stock and barrel to humans. The results of such animal studies are given simplistic labels and their neatly quantifiable findings are extrapolated into the human domain without critical thought. The next move would be to conclude that our behaviour is determined by forces beyond our control.

There is a widely believed fallacy that all of our characteristics are determined by the genes we inherit from our parents. The popularity of books such as Aldous Huxley's *Brave New World* must bear some blame for this. It is strengthened by the discovery that simply inherited chemical defects can cause gross mental retardation (phenylketonuria was the first to be recognized); that Down's syndrome is due most commonly to the presence of an extra chromosome; and that men with two Y chromosomes (instead of the normal one) seemed at one time always to be aggressive criminals. This assumption of genetic determinism, however, has become overstated in the popular mind; genes only *predispose* us to certain characteristics or behaviours, and their expression can very often be modified by changing the environmental conditions (p. 161).

The search for a biological basis for sexual orientation

Scientific interest in the possible biological bases of sexual orientation waxes and wanes. It is often the emergence and application of new techniques which provide fresh ways of tackling enduring questions and spark off a new wave of interest in the topic. Today we are witnessing such a wave of interest, given extra momentum by the 'gay' lobby. In such circumstances, scientific objectivity is put in jeopardy. Until the early 1990s, most research on sexual orientation assumed that homosexuality represents a relative failure of sexual differentiation and that heterosexuality is the norm. More recently, some have argued that homosexuality should not be stigmatized as abnormal. In certain tribal cultures in which homosexuality is expected of all boys before marriage, however, heterosexuality ultimately prevails. This, incidentally, indicates that homosexual behaviour does *not* always or necessarily indicate a homosexual orientation.

In considering the possible biological basis of homosexuality, we follow the lead given by John Bancroft, who listed evidence under four headings: hormonal mechanisms, brain structures, neuropsychological function and genetic factors.

As regards hormonal mechanisms, Bancroft notes that 'the early idea that homosexuals are hormonally different was abandoned some time ago' (1994: 437), as was the assumption that 'rodents provided a model without qualification or justification relevant to the human'. Indeed, the overtly sexual manifestations of gender-specific behaviour in rodents such as lordosis are, says Bancroft, 'clearly unhelpful in studying the human – lordosis has no primate counterpart, and our sexual motor activity cannot be classified in such discreet categories as "mounting"'.

Bancroft notes that Simon LeVay has reported sexual dimorphism in certain brain structures, particularly the nucleus of the pre-optic area of the rat, and that this is smaller in homosexual than heterosexual men); other investigators have focused on possible sexual dimorphisms in the corpus callosum and the anterior commissure. Bancroft concludes that the findings are confusing, since 'there is a lack of either consistency or replication . . . Numbers are inevitably small, and in most studies homosexual subjects have died of AIDS; the possibility that structural changes could be a consequence of disease, such as AIDS, remains. But even if these findings are substantiated, and specific areas of the hypothalamus or elsewhere are found to be linked to sexual orientation, it is difficult to imagine what the nature of such a link would be. It is certainly unlikely that there is any direct relationship between structure of a specific area of the brain and sexual orientation *per se*' (1994: 438).

As regards neuropsychological functions, Bancroft comments on the increased incidence of left-handedness in both female and male homosexuals that 'the proposed hormonal mechanism [*i.e.* increased exposure to androgen during early development] might explain the findings in lesbians; it is difficult to account for this tendency in gay men with a hormonal explanation. Once again, we have a finding which, if substantiated, does not help to explain biological determination of sexual orientation except perhaps indirectly, either as a marker of other relevant processes or via the effects of gender-related behaviour on "sexual development" ' (1994: 438).

It is noteworthy that Bancroft, writing from the viewpoint of a reproductive biologist, and David Myers, from the viewpoint of a psychologist, reach largely similar conclusions about the biological bases of sexual orientation. To Bancroft, 'it remains difficult, on scientific grounds, to avoid the conclusion that the uniquely human phenomenon of sexual orientation is a consequence of multi-factorial developmental processes in which biological factors play a part, but in which psychosocial factors remain crucially important. If so, the moral and political issues must be resolved on other grounds' (1994). Myers concludes: 'Rather than specifying sexual orientation, perhaps biological factors predispose the temperament that influences sexuality in the context of individual learning and experience. Nevertheless, the consistency of the genetic, prenatal, and brain findings has swung the pendulum toward a physiological explanation.' He poses the question: 'If our sexual orientation is indeed something we do not choose and cannot change, then how does the person move towards either a heterosexual or a homosexual orientation?' He believes some consensus has emerged from hundreds of research studies on sexual orientation: (1) homosexuality is not linked with problems in a child's relationship with parents, such as with a domineering mother or/and an ineffectual father, or a possessive mother and a hostile father; (2) homosexuality does not involve a fear or hatred of people of the other gender; (3) sexual orientation is not linked with levels of sex hormones currently in the blood; and (4) sexual victimization of children by an adult homosexual or others is not a factor in homosexual orientation (Myers 1998: 380).

It has been argued that if a 'gay gene' is discovered, this would mean that homosexual orientation was 'natural' and thus homosexual behaviour is not perverted. But an inherited inclination cannot in general establish the rightness or wrongness of any behaviour. John Stott addresses this specifically in expounding Romans 1:26–27:

> Some homosexual people are urging that their relationships cannot be described as 'unnatural', since they are perfectly natural to them. John Boswell has written, for example, that 'the persons Paul condemns are manifestly not homosexual: what he derogates are homosexual acts committed by apparently heterosexual people'. Hence Paul's statement that they 'abandoned' natural relations, and 'exchanged' them for unnatural (26–27) [Boswell 1980: 107ff.]. Richard Hays has written a thorough exegetical rebuttal of this interpretation of Romans 1, however. He provides ample contemporary evidence that the

opposition of 'natural' (*kata physin*) and 'unnatural' (*para physin*) was 'very frequently used . . . as a way of distinguishing between heterosexual and homosexual behaviour' [Hays 1986: 192]. Besides, differentiating between sexual orientation and sexual practice is a modern concept; 'to suggest that Paul intends to condemn homosexual acts only when they are committed by persons who are constitutionally heterosexual is to introduce a distinction entirely foreign to Paul's thought-world', in fact a complete anachronism . . . the only context which he [God] intends for the 'one flesh' experience is heterosexual monogamy, and . . . a homosexual partnership (however loving and committed it may claim to be) is 'against nature' and can never be regarded as a legitimate alternative to marriage (Stott 1994: 77–78).

Personality traits and brain processes

Sexual orientation is one thing, but what about more subtle psychobehavioural tendencies? Do they have an identifiable biological substrate, and, if so, how much control do we have over their emergence? To answer these sorts of questions, behaviour geneticists have depended heavily on twin studies: identical twins afford ready-made subjects for experimenters seeking to tease out the relative influences of hereditary and environment.

Twins developing from a single fertilized egg that has split into two are genetically identical; fraternal twins, developing from separate eggs, are genetically no more alike than ordinary brothers and sisters. Studies involving thousands of pairs of identical and fraternal twins, carried out in such widely different countries as Sweden, Australia and Finland, have come up with consistent findings. In each study, widely used tests of extraversion (outgoingness) and neuroticism (emotional instability) have shown that for both traits identical twins are more alike than fraternal twins; that is, there is a substantial genetic influence on both traits.

It also appears that emotional instability in social relations has a measurable genetic component. A study of 1,500 same-sex, middle-aged twin pairs found that the odds of divorcing if you have a divorced fraternal twin is 1.6 times the normal (the same rate as for those whose parents have divorced). If you are an identical twin and your sibling has divorced, the odds of your divorcing goes up 5.5 times. These and other studies have shifted scientific opinion towards a greater appreciation of genetic influences on behaviour. But their interpretation is complicated. Personality assessments may be carried out only some time after reunions of separated twins, while adoption agencies try to place separated twins in similar homes. Even so, a consensus is emerging which suggests that genetic influences account for almost 50% of person-to-person differences in traits such as extraversion and neuroticism.

There is no doubt that developments in neuroscience are already changing and enriching our understanding of brain and behaviour. There is equally no doubt that if these developments are wisely and compassionately employed, this new knowledge offers enormous potential to reduce human suffering and, to a degree, to increase the overall extent of human happiness. If they become trapped in a reductionist mould or hijacked by a libertarian ideology, however, they are unlikely to benefit individuals or society as much as they might. They must be seen within a wider integrated understanding of how the relationships between the biological, personal and social aspects of living act, so that there is a clear abandonment of a simplistic unidirectional view of the causes of human action.

Accepting that on balance the evidence for a genetic influence on personality traits such as extraversion and neuroticism is broadly correct, what implications may this have for Christian behaviour? Could it be, for example, that those of our Christian brothers and sisters who prefer more boisterous expressions of their faith in their congregational worship are manifesting an outworking of their biologically inherited extraversion, while those who prefer quieter, more private, sedate and perhaps seemingly withdrawn expressions of their faith are manifesting their biological tendency to neuroticism? This is possible, but as fellow Christians we should respect such differences of expression of religiosity, whether in private devotions or public worship, and avoid any suggestion that only *our* preferred way of behaving in worship is *the* biblical way. Those sharing the same credal beliefs may at the same time, for genetic reasons among others, express them differently. This is certainly worth thinking about.

What about the possibility of genetic factors predisposing to one form of sexual orientation rather than another? This is a different issue for the Christian. As regards sexual behaviour, we believe that scriptural teaching is clear in its approval of one sort and disapproval of another. In contrast, the behavioural expression of *shared* doctrinal beliefs may vary according to constitutional differences, provided that all is done is in a seemly manner, decently and in order.

Biopsychosocial influences on spiritual experiences: some salutary case histories

Temporal-lobe epilepsy may produce heightened religiosity or unusual mystical states. For most Christians (who do not suffer from epilepsy), their Christian life may have begun undramatically and yet decisively, and they could not be described as having experienced a sudden religious conversion or any dramatic mystical states. Nevertheless, we are all subject to change. Some of the outstanding men and women of God have recorded marked swings in their awareness of his presence and power. With the benefit of hindsight and informed by the advances in psychiatry, we can be fairly sure that some of these experiences were pathological: some of these God-fearers were obsessive compulsives and some manic depressives; some struggled with specific phobias, and so on.

How does this relate to brain function? The general public, including Christians, are today usually aware that senile dementia is a product of brain deterioration. They see it primarily as a health problem. Notwithstanding, we would all want to assert that the disorganization of psychological processes evident in dementia cannot truly affect one's objective relationship with God, though it may significantly affect one's subjective experience of that relationship. It is at times said that this relationship is maintained through ways and channels that transcend the experience of the body. The experience of Christian believers in the terminal stages of some forms of dementia refutes this. We believe that a biblically based theology can lead to a better, more compassionate understanding of the harrowing spiritual journeys of many of these patients. It can also help us to realize afresh the collective relationship that we and they have in the great hope for them and us – the hope of resurrection in Jesus Christ.

This leads us to consider the general question of linking changed spiritual awareness with neurological state. Several decades ago, British psychiatrist William Sargant speculated about changes in brain processes occurring at the time of sudden religious conversions. Sargant had been impressed by the Russian physiologist Ivan Pavlov's work on classical conditioning, and was intrigued by Pavlov's explanation of the physiological mechanisms underlying such conditioning. Pavlov had claimed to identify four main constitutional temperaments in his experimental dogs; he labelled them, as Hippocrates had done, for the main temperamental types in human beings.

It then became a short step from categorizing an animal as having, for example, a strong excitatory or a strong inhibitory temperament, to identifying what was going on in the brains of these animals as brain inhibition. When these ideas were applied to human behaviour, attention focused on so-called transmarginal inhibition. This, said Sargant, 'once it sets in, can produce three distinguishable phases of abnormal behaviour – the equivalent, paradoxical, and ultra-paradoxical phases. And finally, stresses imposed on the nervous system may result in transmarginal protective inhibition, a state of brain activity which can produce a marked increase in hysterical susceptibility (or, more rarely, extreme counter-suggestibility), so that the individual becomes susceptible to influences in his environment to which he was formerly immune' (1973: 13). Drawing upon these wholly hypothetical brain processes, Sargant attributed some sudden religious conversions to them, generated at some highly charged revival-like evangelistic gatherings.

Sadly for Sargant, his uncritical dependence on Pavlov proved mistaken. The late Professor Oliver Zangwill, a distinguished neuropsychologist, wrote in a review of Sargant's book: 'Few neurophysiologists brought up in the post-Sherringtonian climate have found it possible to take Pavlov's theories – as opposed to his facts – seriously . . . This does not mean that Pavlovian theory is necessarily wrong or lacking in heuristic value. It does however mean that it is esoteric, controversial and hence perhaps an insecure foundation for Dr Sargant's psychological superstructure.' Sargant has undoubtedly assembled an impressive and well-documented survey of ecstatic religious behaviours from around the world; he has not, as an uninformed reader of his books may be forgiven for believing, even begun the task of formulating plausible brain mechanisms underlying such behaviour.

Although Sargant was unsuccessful in his attempts to link neural processes with changed psychological states associated with heightened spiritual awareness, it does not follow that other attempts might not succeed. Explorations of the associations of religiosity and the occurrence of mystical experiences with their possible neural substrates have continued sporadically over the past few decades. The 'religiosity' of epileptic patients has been recognized since the time of Esquirol (1838) and Morel (1860). Early attempts to explain this association favoured an environmental mechanism, noting the frequent tendency for the epileptic to become socially isolated and suggesting that one way of coping was to turn to the consolation that religion may bring. Negative evidence concerning the relationship between epilepsy and religious hallucinations is less often commented on. For example, the classical studies of temporal-lobe epilepsy by Penfield and Jasper (1954) did not report a single case associating religious hallucinations with epilepsy. The same could be said of surveys in the 1950s and 1960s undertaken by a variety of distinguished neurologists. Despite this negative evidence, it is common to attribute the sudden conversion of the apostle Paul to an epileptic attack, noting that Paul fell down and experienced visual and auditory hallucinations with accompanying transient blindness (Acts 9:3–9; 26:12–17). This has led some to argue that Paul's conversion was the result of an epileptic seizure, rather than a mystical experience. Others point out that Paul did not show the subsequent mental deterioration that would normally be associated with a major epileptic episode; they suggest that Paul's hallucinations were understandable because he was an extremely tired traveller, struggling with the heat of the day, and had possibly been neglecting his physical health. Whatever the psychophysiological hypothesis, the fact remains that Paul was faced by the risen Christ and became a changed man.

Kenneth Dewhurst and A. W. Beard (1970) have speculated about the aetiology of the experiences of other notable Christian mystics down the centuries. Robert Thouless (1923) noted that a mood of exultation may stimulate the expression of religious sentiment immediately before an attack of epilepsy. He quotes Dostoevsky's Prince Mishkin in *The Idiot*, who described the experiences of the aura that preceded his epileptic seizures:

> There was a pause just before the fit itself . . . when suddenly in the midst of sadness, spiritual darkness, and a feeling of oppression, there were incidents when it seemed his brain was on fire, and in an extraordinary surge all his vital forces would be intensified. The sense of life, the consciousness of self were multiplied tenfold in these moments, which lasted no longer than a flash of light; all torment, all doubt, all anxieties were relieved at once, resolved in a kind of lofty calm, full of serene, harmonious joy and hope, full of understanding and the knowledge of the ultimate course of things (quoted in Thouless 1923).[1]

David Bear and Paul Feddio, in a quantitative analysis of the behaviour of temporal-lobe epilepsy patients between their fits, reported that those patients where the epileptic focus was in the left hemisphere 'showed a predilection for ideative, contemplative and perhaps verbal expressions of affect as represented in cosmologic or religious conceptualizing' (1977). Bear's analysis of the behaviour exhibited by temporal-lobe epilepsy patients found they had 'a combination of circumstantial concern with details, religious and philosophical interests, and intense, sober affect – a characteristic of virtually every patient, distinguishing them on an individual basis from each of the contrast subjects' (1977). And he continues, 'Thus objects and events shot through with affective colouration may be incorporated into a mystical or religious view of the world' (1977: 87).

Ten years later, Tucker, Novelli and Walker concluded that hyper-religiosity is not a consistent feature of sufferers from temporal-lobe epilepsy, 'although hyper-religiosity and temporal lobe epilepsy may co-occur in a few individuals, it does not appear to be a direct causal relationship between repeated seizure discharge in the temporal lobes and hyper-religiosity' (1987).[2]

We come back to the importance of recognizing the psychobiological unity of the human person. To be aware of this unity, and of the profound effect that constitutional factors may have on our mental and spiritual life, will, we hope, sensitize us to the anguish which some of our Christian friends experience.

In his book *Genius and Grace* (1992), Gaius Davies gives a fascinating account of the lives of John Bunyan, Amy Carmichael, William Cowper, C. S. Lewis, Martin Luther, Gerard Manley Hopkins, J. B. Phillips, Christina

Rossetti and Lord Shaftesbury. Drawing upon historical research and clinical insights, he offers tentative psychiatric diagnoses of the illnesses from which some of these seemed, at times, to have been suffering. In some cases the diagnoses were such that it remains unlikely that there were significant constitutional or endogenous factors at work; in others there can be little doubt that the manifestation of their spiritual distress *did* have a significant biological/biochemical aetiology. These include Luther, Cowper, Shaftesbury and J. B. Phillips.

States of depression certainly affect the spiritual aspects of life. Changed mental states are at times a consequence of inherited or constitutional factors, manifest in biological and/or biochemical abnormalities of the brain or endocrine systems. Clearly, the outworkings of someone's psychobiological unity may exhibit itself in both mental conditions and spiritual distress. Stephen Judge has underlined the possible relation between spiritual states and psychobiological processes: 'Let us suppose, therefore, that every aspect of what it is to be a human person, even a redeemed person with Christ's spirit in us, is embodied in the particular brain that is in our body, and the brain, mind and spirit are not separate entities but different aspects of our identity' (1995). What then do we mean by spiritual depression? His answer is that

If one tries to imagine an independent spirit that is itself depressed, and which therefore fails to provide necessary guidance to the mind and body, this appears to make sense, but we get into difficulties when we start to ask more detailed questions. For example, if it is *only* the spirit that is depressed, and the mind is in good shape, can we not remind ourselves that 'at his hand are pleasures for evermore' and that 'the difficulties of the present situation are as nothing compared with the hope of glory'? If this does not achieve anything for us, then surely it isn't only our spirit that is depressed, we are also in a depressed mental state . . . My proposal is that we regard spiritual depression as a state of being in which the ultimate goals of life (*e.g.* to enjoy him forever) have ceased to connect with our everyday life and practice . . . I submit that to speak of spiritual depression makes little sense unless being in this state has clear-cut consequences in our mental and bodily state, and that we can adequately define spiritual depression in terms of these so-called consequences without appealing to an independent spiritual entity at all (1995).

A balanced view and the way ahead

There are changes in our brains which occur through no choices of our own; Alzheimer's disease is a classic case. On the other hand, what we do to ourselves may affect the workings of the neural substrate of our minds. Increasing self-knowledge has the potential to make us more responsible for our own actions and more attentive to the plight of those who, through no fault of their own, suffer diseases of the mind. Unfortunately, we readily blame the environment for our failures while being all too ready to take credit for our successes. There is surely wisdom in taking seriously some of the research outlined earlier which shows how people are significantly influenced by their biology and their social, environmental and cultural contexts. We ought to regard *ourselves* as agents responsible for our actions, albeit ready to entertain the possibility that *others* have been abnormally subject to their biology and/or outside influences.[3]

Chapter 11

Psychology

Psychology is often said to have explained away religious belief as 'nothing but' wishful thinking and to hold that Christian faith is 'nothing but' a psychological crutch. The 'warfare' metaphor to portray the relation of science to religion seems to have persisted longer between psychology and Christian belief than in some other areas, except perhaps evolutionary theory. How much truth is there in all this? What is current psychology all about? How do psychologists see the similarities and differences between human and animal nature? What about psychotherapy? How should Christians view the many different varieties that are on offer today? Do the findings of social psychologists about our everyday behaviour lend any support to traditional Christian teaching? How may we move towards a constructive partnership between psychology and Christianity?

Psychology is commonly thought to be synonymous with psychoanalysis, psychotherapy and some forms of counselling. It is not just the general public who think this way. Many students embarking upon university and college courses in psychology expect to hear much about Freud and psychoanalysis, and are surprised to discover that in a whole degree course only a few courses are directly relevant to counselling, and only two or three lectures cover Freud, and then for largely historical reasons.

For some Christians, Freud is a bogyman, and psychology is often seen as a threat explaining away some of their most cherished beliefs. The 'warfare' metaphor of the relation of this aspect of science to religion has sadly persisted all too long.

The 'warfare' metaphor in interactions between psychology and religion, past and present

We have already described the unfortunate intrusion of the 'warfare' metaphor into the relation between science and. religion (pp. 26, 31). Although largely discredited by historians of science, it surfaces time after time in discussions of the relation between psychological science and religious beliefs, most markedly with Freud at the beginning of the twentieth century, Skinner in the middle and Crick at the end.

At the end of the nineteenth century there were four significant influences which provided the basis for later studies of the relationship between religion and psychology. These were (1) Francis Galton's studies of the manifestations of religion (*e.g.* prayer); (2) studies of comparative religion and the origins of religion by anthropologists such as James Frazer; (3) the writings of theologians such as W. R. Inge on mysticism and religious experiences; and (4) the beginnings of a systematic psychology of religion (*e.g.* E. G. Starbuck, *The Psychology of Religion*, 1899), culminating in William James's classic, *The Varieties of Religious Experience* (1902).

None of these studies implied a warfare between psychology and religion. Certainly, for William James, the relationship between the two was a strongly positive one, and he sought to explore how psychology could deepen our understanding of the roots and fruits of religion.

As we move into the twentieth century the picture changes. Despite Freud's own disclaimers, he is widely seen as explaining away religious beliefs and exposing religious practice as nothing but the persistence of an interim social neurosis. The fact that his views on the origins of religion have been repeatedly and severely criticized by professional anthropologists, on the ground that many of the 'facts' upon which he based his theories were incorrect, has done little in the popular mind to bring his views into disrepute (see *e.g.* Malinowski 1927; 1936). Freud produced a good story, and it persists long after his views were widely discredited by scholars in related disciplines.

Another major figure of the early part of the twentieth century was Carl Jung, who was both a psychologist and a psychoanalyst. For a time Jung was a close collaborator of Freud, but they came to differ radically in their views of both psychology and religion. For Freud, psychology pointed to religion as a neurosis which could be dispelled and the patient (the human race?) cured, while for Jung religion was an essential human activity, and the task of psychology was to try and understand how human nature reacts to situations normally described as religious. As G. S. Spinks wrote, 'For Freud religion was an obsessional neurosis, and at no time did he modify that judgment. For Jung it was the absence of religion that was the chief cause of adult psychological disorders' (1963).

While Freud and Jung captured the headlines in the psychology and religion interface in the first half of the twentieth century, there were others who wrote on the topic. R. H. Thouless was one of these. His approach was wholly constructive and in complete contrast with the warfare metaphor. He represents a tradition which has continued since the Second World War, with several noteworthy attempts to offer new insights into religion through the eyes of psychology. Notable among these have been G. W. Allport's *The Individual and his Religion* (1951), and Michael Argyle's several books including *Religious Behaviour* and, with Beit-Hallahmi, *The Social Psychology of Religion*. There are thus many excellent books on the psychology of religion which are not infused with any notion of conflict, but, while they are read by psychologists and others interested in deepening our understanding of the part played by religion in our thoughts and feelings, they are not newsworthy because they are unconfrontational. Such, however, was not the case with B. F. Skinner's views of religion.

In the mid-twentieth century, Skinner's views were the most widely publicized of the 'warfare' genre. This was understandable because of his well-deserved reputation as the leading behavioural psychologist of the preceding fifty years. Having achieved considerable success with techniques for shaping and modifying behaviour, Skinner went on to speculate about how such techniques might be harnessed to influence the future of society. He believed that similar principles, based on the effects of rewards and punishments, could explain how religious practice functions psychologically. 'The religious agency', he said, 'is a special form of government under which "good" and "bad" becomes "pious" and "sinful"' (1953: 116). He argued that good things, personified as a god, are reinforcing, whereas the threat of hell is an aversive stimulus; and that both these shape behaviour. Underlying Skinner's approach is a reductionist assumption. He speaks of concepts of god being 'reduced to' what we find positively reinforcing.

Skinner championed the existence of a war between psychology and religion, but an equally distinguished contemporary took a quite different view. Roger Sperry, psychologist, neuroscientist and Nobel Laureate, has criticized the bankruptcy of some forms of behaviourism and accepts the benefits of a positive relationship between psychology and religion as allies engaged in a common task. Sperry's views on religion would sound very strange to conventional Christian believers, but his interpretation of the complementary roles of psychology and religion is important and convincing. Having spent his career studying brain mechanisms, Sperry wrote that he detected among neuroscientists 'a move away from the mechanistic, deterministic and reductionistic doctrines of the pre-1965 science to the more humanistic interpretations of the 1970s' (1988). He argued that it was simplistic to try to reduce humans to 'nothing but' physico-chemical machines. This is an entirely different view from Crick's, despite citing the same evidence. Approaching the end of the twentieth century and drawing heavily upon the work of neuropsychologists, experimental psychologists and cognitive neuroscientists, Francis Crick formulated and publicized his 'astonishing hypothesis': 'The idea that man has a disembodied soul is as unnecessary as the old idea that there was a Life Force. This is in head-on contradiction to the religious beliefs of billions of human beings alive today. How will such a radical change be received?' (1988). His answer was that since (on his view) the idea of the soul is now redundant and discredited by (his interpretation of) neuroscience, religious belief was meaningless.

Defining psychology today

David Myers defines psychology as 'the science of behaviour and mental processes' (1998: 4). In contrast, another psychologist, Howard Gardner, maintains that 'Psychology has *not* added up to an integrated science, and it is unlikely ever to achieve that status. It no longer makes sense to discuss scientific psychology as a tenable long-term goal . . . What does make sense is to recognize important insights that have been achieved by psychologists; to identify the contributions which contemporary psychology can make to disciplines which may some day achieve a firmer scientific status; and finally to determine whether at least parts of psychology might survive as participants in a conversation which obtains across major disciplines' (1992).

The enormous breadth of the subject matter of contemporary psychology leads to a corresponding diversity in the methods and techniques employed by psychologists today. The objectivity and reproducibility of research findings varies depending on whether one is doing single-cell recordings from the superior temporal cortex of an alert monkey looking at pictures of faces, measuring interhemispheric transfer times in normal humans, or studying memory in Alzheimer's patients, racial attitudes in inner-city populations, or cross-cultural differences in ethnic communities – yet all come under the general umbrella of psychology.

It is instructive to compare psychology with biology, which has also diversified into an enormous range of separate disciplines, from molecular biology at one end to ethology at the other, with traditional aspects of morphology and function in the middle and virtually all the sciences basic to medicine (such as microbiology) included along the way. Psychology has a similar span: at one pole very large numbers of researchers are involved in molecular psychology, neuropsychology, physiological psychology, genetic psychology and psychopharmacology; and at the other end of the continuum there are workers in social psychology, which merges into sociology and social anthropology. For this reason alone, it would be unwise to seek a simple formula as to how to relate psychological knowledge to Christian faith. For example, psychologists studying dopamine levels in the nigrostriatal pathways of rats are unlikely to intrude their personal values and beliefs into their data-gathering and interpretation, even though some of those self-same beliefs may have led them to choose such research in the hope that it might ultimately help those smitten by Parkinson's disease. In contrast, the possibility that personal beliefs may influence professional practice becomes acute for clinical psychologists in, for example, family therapy. The therapist may have to navigate a narrow path between violation of professional commitment in the therapist-client relationship, and commitment to personal beliefs and values.

The scientific status of psychology is vigorously debated by those who espouse the 'postmodernist' cause. Such people question the objectivity of all science, and emphasize the ubiquitous, indeed (they would say), pervasive influence of the subjective. They claim that psychology should aim to 'understand' people, and assert that this is impossible within the traditional objective constraint of experimental science. Such a position further contends that in *qualitative* investigations it is legitimate to approach one's research with a frank political agenda, such as feminism. This contrasts with the traditional assumption that scientists have a duty to keep their prejudices and personal opinions under control when describing nature. Indeed, it strikes at the very roots of science.

Paul Gross and Norman Levitt have highlighted this conflict by focusing on the nature of the scientific enterprise:

> Science is, above all else, a reality-driven enterprise. Every active investigator is inescapably aware of this. It creates the pain as well as much of the delight of research. Reality is the overseer at one's shoulder, ready to rap one's knuckles or to spring the trap into which one has been led by overconfidence, or by a too-complacent reliance on mere surmise. Science succeeds precisely because it has accepted a bargain in which even the boldest imagination stands hostage to reality. Reality is the unrelenting angel with whom scientists have agreed to wrestle (1994: 234).

Michael Morgan has underlined this: 'The claim that data are always corrupted by opinion is, frankly, bogus. Even if it were partly true, that would be all the more reason to redouble our efforts to be objective.'

Psychology and the relevance of faith

In the early years of the twentieth century the lay person could be forgiven for believing that psychology was synonymous with Freudian psychoanalysis, at mid-century that it was equivalent to Skinner's behaviourism, and today that psychology is merely another name for psychotherapy and counselling. That person would be wrong.

What constitutes psychology today? It contains at least the following ingredients: (1) basic sensory processes whereby we gather information from the world we live in: the senses of sight, hearing, taste, touch, smell; (2) the way in which we process information so gathered through the senses; (3) basic processes of conditioning and learning; which in turn are complemented by (4) a large effort investigating how we store the information so acquired, in other words the study of memory; the latter also requires us to devote considerable energies to (5) understanding how we retrieve stored information and gain access to it as and when we need it. And then there is the further area, greatly influenced today by artificial intelligence and computer studies, of (6) our attempts to understand complex processing, traditionally described as thinking. Even when such information has been gathered there remains (7) a large area of psychological enquiry which devotes itself to understanding how we organize and execute the motor responses we make when we have successfully gathered and processed information from our environment. The final ingredient is (8) understanding the basic processes of emotion and motivation.

In all of these diverse areas, psychologists, like other scientists, have developed models and proposed theories, sometimes expressed mathematically, often formulated in terms of information flow diagrams, in order to understand the mechanisms at work and to enable predictions to be made about what will happen if certain specific kinds of further experiments are carried out. This kind of work seems to occupy at least a third of the total field of modern psychology. A further one third is contributed by parallel studies, also carried out by psychologists, seeking to understand the biological bases undergirding the psychological processes already listed.

There remains the last third, made up of how the development of all the above processes takes place from birth through to death; how we relate to one another in social situations; and how we can best understand and describe human personality.

It should already be evident that when we talk about the integration of Christian belief and psychology, it cannot mean that, in the study of (say) sensation and perception or of depth perception and stereopsis, it matters whether the experimenter is a Christian or a non-Christian, or whether we are using as our experimental subjects those who hold Christian beliefs or those who do not. Likewise, we do not expect the principles of classical conditioning to differ between Christians and non-Christians, nor do we expect the schedules of reinforcement found to apply in operant conditioning to differ between Christians and non-Christians. In the case of memory, it is difficult to think of any reason why the information-processing models currently fashionable will differ between Christians and non-Christians. Presumably the serial-position effect in learning word-lists will be the same for Christians and non-Christians. We have no grounds, on the basis of Christian belief, for selectively modifying the distinction usefully made between declarative memory and procedural memory. Any attempt to *integrate* or *incorporate* Christian beliefs with psychological theorizing in these areas of contemporary psychology would clearly be misplaced, or bluntly 'nonsense'.

Likewise, we can ask whether we have any grounds for believing that, for example, the stages in the normal motor development of a child are likely to differ between Christians and non-Christians. Presumably we do not believe that a new theory of motor development should be put forward which 'incorporates' Christian beliefs with the empirical evidence. Neither do we expect the milestones of cognitive development to differ as between Christians and non-Christians. Any attempt to mix Christian beliefs with psychological accounts is clear evidence of a failure to recognize the different domains to which the two kinds of knowledge belong and the different categories they use for expressing that knowledge; it is guaranteed to cause confusion and to make nonsense of both.

Where relating psychology to Christian beliefs becomes problematic is with personality theory, and its applied aspects in clinical psychology and psychiatry. These have occupied roughly 10% of the reports in the *Annual Review of Psychology* since the late 1950s. Although this is a relatively small part of contemporary psychology, it is here that the liveliest debates about how properly to relate psychology and Christian belief occur. There are good reasons for this. Although personality theory, or personology as it is now called, is a minor part of the subject of psychology, nevertheless it comprises a large proportion of the applications of contemporary psychology, particularly by clinical psychologists and psychotherapists. It has to be addressed in any discussion of the relation of psychology and Christian belief.

To eschew incorporating Christian beliefs with psychological theories does *not* prevent us from identifying recurring themes therein which may conflict with our Christian understanding of human nature. After all, the Bible is full of profound insights into human nature, both individually and in society. There are several pervasive themes in contemporary psychology which may challenge some of our basic Christian beliefs and potentially sow the seeds of conflict.

As we saw in the last chapter, psychologists devote much effort to understanding basic cognitive mechanisms and to studying the biological substrates of those processes. Here the accumulating evidence points to an ever-tightening link between mental processes and their biological substrates. We have also noted how contemporary neuropsychology seems to demand a rethinking of the nature of humankind. A second issue arises from the widespread use by psychologists of animals to help our understanding of human cognition and behaviour. As this work has developed, mind-like behaviour in non-human primates has been studied increasingly, and raises questions of the relation of human nature to animal nature. Do animals possess souls? Does the spirit uniquely separate humankind from animals? There are also issues that recur as we consider what personologists have to say about human nature. How should we relate their theories to traditional Christian views?

Human nature and animal nature

For obvious and good ethical reasons, our understanding of changes in brain structure on human mental life and behaviour depends on studying brain damage due to genetic factors or other influences, rather than invasive experimental procedures. While the techniques available for locating brain damage in patients have made enormous leaps in the last three decades, there remains a measure of uncertainty about the precise location and extent of any damage, which can be resolved only at post-mortem examination. This is one reason why work on animals is so important. With careful surgery it is possible to produce localized changes in the brain and then study differences between pre-operative and post-operative performance on a variety of experimental tasks given to the animals. Such studies help to reduce the uncertainties which remain if we are able to study only accidentally brain-injured people.

There is a tacit assumption that animal studies, while interesting in themselves, throw light on human mind–brain links. This in turn depends on further assumptions about similarities between the brains of humans and animals. This is not new in psychology. For many decades attempts to discover more about human conditioning and learning have been driven by carefully controlled studies on animals. Significant examples are Pavlov's work on classical conditioning using dogs, Kohler's research with non-human primates, and Skinner's studies of classical and operant conditioning in rats. The question naturally arises: what implications do they have for our understanding of human, as distinct from animal, nature?

The traditional interests of comparative psychologists in animal behaviour have been extended by ethologists, primatologists and evolutionary psychologists. They have investigated, under more controlled conditions than is possible with humans, such processes as learning, perceiving and remembering, and also how animals from different phyla and with nervous systems of increasing complexity reflect this in changes in behaviour, both instinctive and learned. The brilliant work of ethologists like Konrad Lorenz, Karl von Frisch and William Thorpe have helped psychologists break free from the straitjacket imposed by earlier generations who focused rather narrowly on conditioning, maze learning, escape from puzzle boxes, and the like.

Not surprisingly, psychologists with a primary interest in how brain processes mediate human cognition and behaviour continue to use animals widely in their research. It is important to recognize that those who work in these areas, while often sharing common research objectives, scientific training and skills, are as diverse in their personal metaphysical beliefs, ideologies and hidden agendas as any other group of people. When it comes to interpreting their results, they sometimes differ widely: if they start as materialist reductionists they will present the human being as 'nothing but' an exceptionally complex primate; others, while not adopting a Christian position, will still write readily about the 'uniqueness' of humankind, even though attributing such uniqueness to different aspects of human cognition and behaviour. Yet others, starting from Christian presuppositions, will recognize the many similarities of humans with non-human primates while noting significant differences between them.

The picture emerging

What then is the overall picture? From comparative psychology, two things stand out.

The first concerns language. In their different ways, Robert Hinde (1987) and William Thorpe (1974) both identified this as one of the crucial distinguishing features of humanness. Hinde regards the key difference as one of quality rather than quantity. We know from the work of David Premack and others that chimpanzees are able to use forms of language. The important factor is that our capacity is so much greater than animals that it can be meaningfully thought of as qualitatively different. This is not universally agreed. Fooks, for example, has argued that the possibility of chimpanzee language challenges the notion of a sharp break between human and animal cognition. Richard Byrne would endorse a similar view at least as regards 'reading the minds of others' (1995). In contrast, Thorpe and Hinde believe that there is a point where 'more' becomes a discreet difference. Coming at the same issue from the point of view of a neuropsychologist and of a comparative anatomist respectively, both Ettlinger and Passingham conclude that it is aspects of language such as 'the ability to represent information *within* an individual's mind' (Ettlinger 1984) and 'our ability to represent words to ourselves; to manipulate internal systems' (Passingham 1982) that form the crucial difference between humans and non-human primates.

In their work with pigmy chimp 'Panzi', Duane Rumbaugh and Sue Savage-Rumbaugh suggest that a chimp raised from infancy in a language-rich environment can equal the language comprehension and (less convincingly) non-vocal language expressions of a three-to-four-year-old human child – and that is impressive. In multiple cognitive dimensions, however, the abilities of the human three-year-old seem to be a ceiling for the chimp. Such data may point on the one hand to a continuity in fundamental abilities and, on the other, to something unique emerging in the interactive play of similar but substantially enhanced cognitive abilities in the human. Is this the 'phase change' discussed by Polkinghorne (see p. 257)? Or, at least, part of it?

Secondly, in his Gifford Lectures (1974), Thorpe reviewed a vast amount of empirical evidence, and made the very strong claim that we are left with a tremendous chasm – intellectual, artistic, technical, linguistic, moral, ethical, scientific and spiritual – between ape and human. He maintained that as scientists we have no clear idea how this gap is bridged. Humankind is unique in many aspects, and we may never know, scientifically, how this happened. In the light of more recent work by Byrne and others, we would qualify this slightly, and argue for the possibility of what we can legitimately label ethical behaviour in the way that chimpanzees look after their young and conduct their family life. But there still remains a vast gulf in scientific, literary, artistic, cultural and technological achievements between humankind and all other animals. It is possible to be enthusiastic about the results of ongoing research on non-human primates, whether in the field or the laboratory, and to note some of their hitherto underestimated cognitive abilities, but not to go on to believe naïvely that therefore they are little different from ourselves. What is surprising is that chimpanzees, despite being genetically so similar, are so vastly different from us: no libraries, no art galleries, no technology, no science, no symphony orchestras, no religion. While so similar to us in both genes and brain structure, we do not know how or when the quantum leap occurred which made possible all the uniquely human achievements. As regards the spiritual dimension, it is true that animals can be made to exhibit 'superstitious' behaviour, as B. F. Skinner did with pigeons, but that is hardly what we normally mean by spiritual!

In some of this there is a lesson for those of us who, as Christians, wish to assert and defend the uniqueness of humankind. Although it would be dangerous to assert that there are no structures or fundamental cognitive processes unique to the human brain, it would be more helpful as well as in accord with the evidence to note that the neural substrate of *Homo sapiens* confers a language capacity for internal representation and symbol manipulation of an order far beyond anything observed in non-human primates. What the final story is on this issue remains to be seen. We have no religious stakes in the outcome. No fundamental Christian doctrine or biblical understanding rests on the cognitive uniqueness of humans.

A Christian perspective

It is salutary to recall that some of the issues raised in 1902 by A. M. Fairbairn in his book *The Philosophy of the Christian Religion* are the same today. He wrote concerning the position of man and his nature: 'Do the eloquently minimized differences which we find in the structure of man, as distinguished from the man-like ape, explain the differences in their histories?' We have to accept that we are still puzzled why such seemingly small differences between humankind and the non-human primates have given rise to such vastly different outcomes in art, literature, science, music, religion, technology and so on – the 'vast gap' described by Thorpe (1974). This same point has been strongly emphasized by Professor Steven Rose:

> Great disservice has been done to biology by the contingencies of the historical development of our subject which meant that in the early years of this century two separate sub-disciplines emerged: genetics, which essentially asks questions about the origins of *differences* between organisms, and developmental biology, which asks questions about the processes which ensure *similarity*. The careless language of DNA and molecular genetics serves to widen this gap rather than help bridge it so as to open the route towards the synthetic biology that we so badly need. Let me state the biological problem in its bluntest possible terms: chimpanzees and humans share upwards of 98% of their DNA, yet no-one would mistake a chimpanzee phenotype for a human phenotype. We have no idea at present about the developmental rules which lead in one case to the chimp, in the other to the human, yet this, surely one of the great unsolved riddles of biology, seems a matter of sublime indifference to most molecularly orientated geneticists (1992).

Nothing is to be gained scientifically or theologically by glossing over the real differences between humankind and animals. Fudging the issue of the far-reaching *differences* conferred by, for example, some aspects of human language with high-sounding sentimental slogans, serves the best interests of neither animals nor humans. We simply 'deceive ourselves and the truth is not in us' (1 Jn. 1:8). At the same time, there are no theological issues at stake in fully recognizing the many and equally important *similarities* between humans and animals. Scientific and medical research has much to gain by recognizing and building upon these similarities. We can find no biblical reason for denying that animals have conscious experience and distinct mental abilities. As we have seen, the term *nepeš* or 'living being' applies in Scripture equally to animals and humans (p. 140). But we need to beware of extrapolating this too far. An analogy may help here:

> . . . a weak mixture of gas and air may contain the same kinds of molecules and lie on the same continuum as a richer mixture that burns, but that does nothing to prove that some kind of flame must be possible in the weaker mixture. Below a certain minimum concentration the mixture is simply inflammable. By the same token, the fact that a human brain organized in a specific way can embody conscious mental and spiritual life does nothing to prove that similar mental and spiritual capacities must be present in the brains of animals. If the biblical claim were that man is distinguished from lower animals . . . by having a brain sensitive to non-physical influences, and [if] these non-physical influences are what make his behaviour essentially human, then of course the question of the capacities of animals . . . would become crucial. Once this claim is recognized as without biblical foundation, however, the polemical pressure disappears, and the whole issue becomes of marginal interest to the biblical apologist, whose primary duty regarding it is to keep an open mind (MacKay 1988: 73).

Psychology, psychotherapy and the humanizers of science

A Christian psychotherapist meeting with a client who is self-evidently suffering mental anguish does not have the option (one might say the luxury) of the research psychologist of critically reflecting on the ultimate scientific basis of a personality theory by which he will be guided in judging how best to help his client.

The problems of controlled experimentation in areas of research such as cognitive psychology, psychophysics, learning theory and psychobiology fade into insignificance compared with the problems confronted by those psychologists seeking to carry out research aimed at understanding personality. Almost any single aspect of personality involves biological, cognitive, social, environmental as well as general psychodynamic variables. All are relevant and all have interest to the researcher. This daunting complexity helps to explain why there are more than 250 types of psychotherapy on offer in the psychological marketplace at the present time.

Where theories are so difficult to formulate and are therefore inevitably tentative, the scope for importing personal values and beliefs becomes great. It is for this reason that issues around psychotherapy are of particular importance to Christians. Different knowledge domains come together in the theory and practice of psychotherapy. We recognize several relevant to the therapist, including the theory of (normal) personality and development, and theories of psychopathology. Moreover, any comprehensive theory of psychotherapy must include also an understanding of the processes of change.

In the present discussion, we shall deliberately confine ourselves to the theory and practice of psychotherapy today, which are matters of increasing concern among Christians. Though related to psychiatry and to general clinical psychology, psychotherapy is distinguishable from those areas of research and practice. In deciding how best to deal with such a vast topic, we were helped at the time of writing by a book on psychology and psychotherapy by Robyn Dawes, Professor of Psychology at one of America's most distinguished universities, Carnegie Mellon University, and held in high esteem by his peers in North America, as evidenced by the award in 1990 of the American Psychological Association's William James Award. In a review of the book, *House of Cards: Psychology and Psychotherapy Built on Myth* (1994), Gregory Miller emphasized a point relevant here when he wrote: 'This is not a book by a hostile outsider attacking things of which he is ignorant. It will surprise many readers in various camps that Dawes concludes unequivocally, "psychotherapy works overall in reducing psychologically painful and often debilitating symptoms" (p. 38).' Miller's general conclusion about Dawes's book was that 'This is a major work, strong in its science, impressive in its scope of issues and examples treated knowledgeably, and bold in its policy statements'. We could add as a footnote that Dawes's book was endorsed by one of the leading American psychologists involved in the evaluation and criticism of the recovered-memories debate. Elizabeth Loftus comments: 'Robyn Dawes has skillfully and disturbingly revealed the clay feet of some of psychotherapy's deepest "truths". In so doing, Dawes has created a giant of a book – giant in its vision, giant in its accomplishments, giant in its implications' (1993). We regard Dawes's work as worthy of careful attention, because the whole issue of psychotherapy is such an extremely sensitive one today, in both Christian and non-Christian circles.

Students often choose to study psychology because they see it as a means to an end. They want to do something to help other people. Such a motive is, we suspect, especially powerful among Christian students. They see it as a way of following their Master, who, by teaching and example, showed care and compassion to all in distress. It thus becomes a natural step to view the study of psychology as opening the way to a career as a psychotherapist or a counsellor. There, they assume, they will be ideally placed to fulfil their Christian calling.

A Christian embarking on a career as a psychotherapist has to face many serious issues.[1] For example, some of the personality theories on which the practice of psychotherapy is based depend on undisclosed and often un-Christian presuppositions about what life is all about, such as what should be the primary goals in life and what is the highest good to which humankind should aspire.

Paul Vitz (1977) has exposed the anti-religious assumptions of some of these secular psychologies. These include atheism or agnosticism, naturalism, reductionism, individualism, relativism, subjectivism and gnosticism (knowledgism). Allan Bergin (1980) has added criticisms of humanistic psychology, such as that people are fundamentally good and not disposed to evil, that to be true to oneself is the highest good, and that self-analysis can reveal more important truths than are revealed by scientific analysis. Mary Stewart van Leeuwen comments: 'Neither naturalism or evolutionism will suffice as the foundational anthropology for the Christian psychologist . . . Central to the Christian's understanding of human nature is the conviction regarding the person's ongoing relationship with God, a relationship that was intended to be of both a providential and a covenant sort' (1982: 49).

The need to expose such hidden assumptions and undeclared metaphysical aims is a concern not exclusive to Christians. The intrusion of non-psychological influences into psychotherapy is well illustrated by the influence of the New Age movement. Dawes draws attention to a common New Age emphasis:

> Poor self-esteem is often cited as the 'root cause' for everything from a failure to learn in elementary schools, to failure in business, to 'over-achievement', to divorce, or even to 'sexual co-dependency'. . . Let me state categorically that there is *no* 'scientific evidence' that people who have deep insecurities and self-doubts have nothing to contribute to the world. The most casual reading of biographies indicates that many admirable people, like Abraham Lincoln, suffer from 'deep insecurities and self-doubts', and that many less than admirable people suffer no self-doubts whatsoever, at least until they were caught or disgraced . . .
> The 'scientific evidence' to which Branden [a typical author on this theme] refers is a *correlation* between the level of self-esteem and the degree to which people engage in personally and socially positive activities and avoid negative ones. My great-grandmother would not be surprised by the existence of this correlation. Of course, she might say it's because people who do rotten things feel bad about

themselves and people who do good things feel good about themselves. She might add, damned good thing too, or society would fall apart (1994: 234).

Dawes goes on to point out that this obsession with self-esteem to the exclusion of almost everything else 'discourages trying to change one's behaviour or life course. Instead, it encourages shoring up self-esteem first by running to a therapist or group' (1994: 243).

His cynical conclusion is that

> professionals in psychology and psychotherapy clearly benefit from a New Age Psychology – it brings them clients. Unfortunately, they in turn contribute to and reinforce that psychology. Having lost sight of scientific skepticism and the need for careful research, the 'professionals' view' has become highly compatible with the New Age view . . . without adherence to the scientific standard of 'show me', professional psychology and psychotherapy become a matter of 'views' and 'schools', with the result that they are highly influenced by cultural beliefs and fads: currently, the obsession is with 'me' (1994: 250).

Mary Stewart van Leeuwen has drawn attention to the danger of selecting any one of several personality theories currently on offer in the psychological marketplace and seeking to baptize it with Christian orthodoxy (1982).[2]

Dawes has also highlighted an issue which our current preoccupation with individualism may have caused us to neglect. He notes that 'emotional suffering is very real, and the vast majority of people in these expanding professions sincerely wish to help those suffering. But are they really the experts they claim to be . . .? Are they better therapists than minimally trained people who may share their knowledge of behavioural techniques, who are empathetic and understanding of others?' (1994). He points out that 'these questions had been studied quite extensively, often by psychologists themselves. There is now an impressive body of research evidence indicating that the answer to these questions is no' (1994). If some of the claims of psychotherapy are not demonstrably true, and/or if the theories of personality undergirding the practice of psychotherapy cannot be defended from relevant evidence, then the Christian psychotherapist must be the first to question any false claims. Here is an issue where Christians ought to stand out as radical reformers, fulfilling our functions as salt and light and exposing error in a fallen world.

Having said this, let us reiterate (in order to avoid any possible misunderstanding) that we are not here discussing serious mental disorders which may require psychopharmaceutical intervention, expert cognitive or behavioural therapy or other extensive therapeutic interventions. We are discussing the kind of therapy offered to what may be described as the mildly unhappy person. Great skill and extensive training are required to diagnose and assess suitable treatment for the malaise, and here a well-trained therapist is needed rather than a lay counsellor. None of this, however, invalidates one of Dawes's major points, namely that the theories and practices of the well-trained therapist must continuously and relentlessly be put under the critical scientific microscope. Values and beliefs smuggled in and presented as part and parcel of the science must be ruthlessly exposed. Our Christian valuation of human nature demands nothing less.

For Christians, there is a further issue concerning the stewardship of our resources. If Dawes is right,

> . . . we should not be pouring our resources and money to support high-priced people who do not help others better than those with far less training would, and whose judgments and predictions are actually worse than the simply statistical conclusions based on 'obvious variables'. Instead we should take seriously the findings that the effectiveness of therapy is unrelated to the training and credentials of the therapist . . . in attempting to alleviate psychological suffering, we should rely much more than we do on scientifically sound community-based programmes and on 'para-professionals' who can have extensive contact with those suffering at no greater expense than is currently incurred by paying those claiming to be experts (Dawes 1994).

In the context of the Christian faith, have we lost or underplayed some of the benefits of mutual support and understanding clearly taught in Scripture as intrinsic features of the local church, the community of believers? Have we failed to mobilize the considerable resources of gifted 'para-professionals' at the expense of focusing too much on a small, select band of so-called professional psychotherapists?

Social psychology and Christian belief

Some Christian psychologists link their profession with their faith by viewing religious phenomena, such as conversion or prayer, through a psychological microscope (an exercise often called 'the psychology of religion'). Others study psychology *and* religion by correlating personality theories with theological assumptions or by proposing Christian approaches to counselling. Still others relate psychology and religion by asking how major insights into human nature from psychological research compare with biblical and theological ideas about human nature. We have already seen how research in neuropsychology underlines the unity of the human person and have noted how this has been a major theme in the writings of biblical scholars in the twentieth century. We now ask: how well do biblical assumptions about human nature connect with the human image emerging from research in social psychology (the scientific study of how people think about, influence, and relate to one another)?

David Myers (1994) has identified four two-sided truths which constitute what he calls social psychology's 'big ideas'. Myers's 'big ideas' in a specifically Christian context are as follows.

Rationality and irrationality

How 'noble is reason' and 'infinite in faculties' is the human intellect! Thus rhapsodized Shakespeare's Hamlet. In some ways, indeed, *our cognitive capacities are awesome*. The 1.5 kg tissue in our skulls contains circuitry more complex than all the telephone networks on the planet, enabling us to process information either laboriously or automatically, to remember vast quantities of information, and to make snap judgments using rule-of-thumb heuristics. One of the most human of tendencies is our urge to explain behaviour, to attribute it to some cause, and therefore to make it seem orderly, predictable and controllable. As intuitive scientists, we make our attributions efficiently and with enough accuracy for our daily needs.

Such views are echoed by Jewish and Christian theologians. We are *made in the divine image* and given stewardship for the Earth and its creatures. We are the summit of the creator's work, God's own children.

Yet, in ways of which we are often unaware, *our explanations and social judgments are vulnerable to error*, insist social psychologists. When observing others we may be biased by our preconceptions and 'see' illusory relationships and causes, treat people in ways that trigger their fulfilling our expectations, be swayed more by vivid anecdotes than by statistical reality, and attribute others' behaviour to their dispositions (for example, by assuming that someone who acts strangely must *be* strange). Failing to recognize such sources of error in our social thinking, we are prone to overconfidence in our social judgments.

Such conclusions have a familiar ring to theologians, who remind us that *we are finite creatures* of the one who declares 'I am God, and there is none like me' (Is. 46:9), and, 'As the heavens are higher than the earth, so are my ways higher than your ways and my thoughts than your thoughts' (Is. 55:9). As God's children we have dignity, but not deity. Thus we must be sceptical of those who claim for themselves god-like powers of omniscience (reading others' minds, foretelling the future), omnipresence (viewing happenings in remote locations), and omnipotence (creating or altering physical reality with mental power). We should be wary even of those who idolize their religion, presuming their doctrinal fine points to be absolute truth. Always, we see reality through a dim mirror (1 Cor. 13:12).

Self-serving bias and self-esteem

Our views of ourselves are brittle containers of truth. Heeding the admonition to 'know thyself', we analyse our behaviour, but hardly impartially. Our human tendency to *self-serving bias* appears in our differing justifications for our successes and failures, for our good deeds and bad. On socially desirable criteria, we commonly view ourselves as relatively superior – as, say, more ethical, socially skilled and tolerant than our average peer. Moreover, we excuse our past behaviours; we have an inflated confidence in the accuracy of our beliefs; we misremember our own past in self-enhancing ways; and we overestimate how virtuously we would behave in situations that draw less-than-virtuous behaviour out of most people. Anthony Greenwald speaks for ordinary men and women as well as for scientists: 'People experience life through a self-centred filter' (1984).

That conclusion echoes a very old religious idea – that *self-righteous pride is the fundamental sin*, the deadliest of the seven deadly sins. Thus the psalmist could declare that 'no-one can see his own errors' (Ps. 19:12), and the Pharisee could thank God 'that I am not like other men' (Lk. 18:11; and, of course, you and I can thank God that we are not like the Pharisee!). Pride goes before a fall. It corrodes our relations with one another, as in conflicts between partners in marriage, management and labour, nations at war. Each side views its motives alone as pure, its actions beyond reproach. But so does its opposition, continuing the conflict.

Yet *self-esteem pays dividends*. Self-affirmation is often adaptive. It helps maintain our confidence and minimize our depression. To doubt our efficacy and to blame ourselves for our failures is a recipe for failure, loneliness or dejection. People made to feel secure and valued exhibit less prejudice and contempt for others.

Again, there is a Christian parallel, in the idea that to sense an *ultimate acceptance* (divine 'grace' – the theological parallel to psychology's 'unconditional positive regard') is to be liberated from both self-protective pride and self-condemnation. To feel profoundly affirmed, just as I am, lessens my need to define my self-worth in terms of achievements, prestige, or material and physical well-being. It is rather like the insecure puppet Pinocchio saying to his maker Geppetto, 'Papa, I am not sure who I am. But if I'm all right with you, then I guess I'm all right with me.'

Attitudes and behaviour

Studies during the 1960s shocked social psychologists with revelations that our attitudes sometimes lie dormant, overwhelmed by other influences. But follow-up research was reassuring. *Our attitudes influence our behaviour* – when they are relevant and brought to mind. Thus our political attitudes influence our behaviour in the voting booth. Our smoking attitudes influence our susceptibility to peer pressures to smoke. Our attitudes toward famine victims influence our contributions. Change the way people think and – whether we call such persuasion 'education' or 'propaganda' – the impact may be considerable.

Social psychology teaches us that this may work either way: we are as likely to act ourselves into a way of thinking as to think ourselves into action. We are as likely to believe in what we have stood up for as to stand up for what we believe. Especially when we feel responsible for how we have acted, *our attitudes follow our behaviour*. This self-persuasion enables all sorts of people – political campaigners, lovers, even terrorists – to believe more strongly in that for which they have witnessed or suffered.

The realization that inner attitude and outer behaviour, like chicken and egg, generate one another parallels the Jewish-Christian idea that inner faith and outer action likewise feed one another. Thus, *faith is a source of action*. Elijah is overwhelmed by the Holy as he huddles in a cave. Paul is converted on the Damascus road. Ezekiel, Isaiah and Jeremiah undergo an inner transformation. In each new case, a new spiritual consciousness produces a new pattern of behaviour.

But *faith is also a consequence of action*. Throughout the Old and New Testaments, faith is seen as nurtured by obedient action. Philosophers and theologians note how faith grows as people act on what little faith they have. Rather than insist that people believe before they pray, Talmudic scholars would tell rabbis to get them to pray and their belief will grow. 'The proof of Christianity really consists in "following",' declared Søren Kierkegaard (1851). To attain faith, said Pascal (1670), 'follow the way by which [the committed] began; by acting as if they believed, taking the holy water, having masses said, *etc*. Even this will naturally make you believe . . .' C. S. Lewis concurred:

> Believe in God and you will have to face hours when it seems *obvious* that this material world is the only reality; disbelieve in Him and you must face hours when this material world seems to shout at you that it is not all. No conviction, religious or irreligious, will, of itself, end once and for all [these doubts] in the soul. Only the practice of Faith resulting in the habit of Faith will gradually do that (1960).

Persons and situations

Myers's final two-sided truth is that people and situations interact. We see this, first, in the evidence that social influences powerfully affect our behaviour. Studies of conformity, role-playing, persuasion and group influence do not, in general, present a flattering picture of humankind. The most dramatic findings come from experiments that put well-intentioned people in evil situations to see whether good or evil prevailed. To a dismaying extent, evil pressures overwhelm good intentions, inducing people to conform to falsehoods or capitulate to cruelty. Faced with a powerful situation, nice people often do not behave nicely. Depending on the social context, most of us are capable of acting kindly or brutally, independently or submissively, wisely or foolishly. In one irony-laden experiment, a group of theological students on their way to giving a talk on the parable of the good Samaritan failed to stop and give aid to a slumped, groaning person – *if* they had been pressed to hurry (Darley & Batson 1973). External social forces shape our social behaviour.

The social-psychological idea that there are powers greater than the individual is paralleled by the biblical ideas of *transcendent good and evil powers*, symbolized in the creation story as a seductive demonic force. Evil involves not only individual rotten apples here and there. It also is a product of 'principalities and powers' – corrosive forces – that can make a whole barrel of apples go bad. And because evil is collective as well as personal, responding to it takes a communal religious life.

Although powerful situations may override people's individual dispositions, social psychologists do not view humans as mere thistledown, blown this way and that by social winds. Different people may react differently to the same situation, depending on their personality and culture. Feeling coerced by blatant pressure, they will sometimes respond in ways that restore their good sense of freedom. A minority group may sometimes oppose and sway the majority. When they believe in themselves, maintaining an 'internal locus of control', they sometimes work wonders. Moreover, people choose their situations – their college environments, their jobs, their locales. And their social expectations are sometimes self-fulfilling, as when we expect someone to be warm or hostile and they become so. In such ways, *we are the creators of our social worlds.*

To Christian believers, that rings true. *We are morally responsible* – accountable for how we use whatever freedom we have. What we decide matters. The stream of causation from past to future runs through our choices.

Faced with these pairs of complementary ideas, framed either psychologically or theologically, we are like someone stranded in a deep well with the two ends of a rope dangling down. If we grab either one alone we fall deeper. Only when we hold both ropes can we climb out, because at the top, often beyond where we can see, they come together around a pulley. Grabbing only the rope of rationality or of irrationality, of self-serving pride or of self-esteem, of 'attitudes first' or of 'behaviour first', of personal or of situational causation, plunges us to the bottom of a well. So instead we have to grab both ropes, perhaps without yet fully understanding how they relate. In doing so, we may be comforted that in both science and religion a confused acceptance of complementary principles is sometimes more honest than an over-simplified theory that ignores half the evidence.

The continuing search for a constructive partnership

There is a widespread impression in some quarters that psychology has 'explained away' religious experience and behaviour and that religious beliefs are 'nothing but' wishful thinking. It is both irrational and wrong. Two of the twentieth century's major figures in academic and applied psychology, whose enduring contributions are increasingly acknowledged, held explicit, sympathetic and constructive views of religion. Both Gordon Allport in the USA and Frederic Bartlett in Britain made a point of emphasizing the potential for a positive co-operative relationship between psychology and religion, at the same time underlining the limits of psychological enquiry, at least when practised as a science.[3]

Both Allport and Barlett were wholly convinced of the potential benefits of psychology. They also recognized the distinctive approaches to the gaining of knowledge possible through the scientific enterprise, a view already well articulated by leading physical scientists of earlier generations, as we have seen repeatedly.

Christians and psychology

Just as there are psychologists antagonistic to religion, so also there are Christians who are antagonistic to psychology. Hendrike Vande Kemp has noted that

... such anti-psychologists seem to regard psychology as offering alternative answers to the same questions answered by Christian theology and biblical revelation, questions concerning knowledge of God and salvation history and a proper human response to both. Psychologists, for the most part, are not interested in 'knowing God'. They are interested in what kinds of images of God persons entertain and what beliefs they embrace, and how their faith relates to practice – but these involve 'knowledge' of a very different sort ... the most conservative of the anti-psychologists, who reject all sources of knowledge other than authority, should be equally sceptical of empiricism (or science), rationalism (or philosophy), and mysticism (or phenomenology), as all involve 'excessive curiosity'. Since no form of psychology would be acceptable to them, there is little point in presenting an argument. One might challenge them, however, as to their exegetical method – it is hard to envision one that involves neither induction, deduction nor intuition (1987).[4]

Herman and Marco Aquinis

. . . believe that to posit a mutually beneficial relationship between psychological science and religion, one must address the more fundamental and core features pertaining to the essential nature of the scientific method and religion . . . Fundamental differences between science and religion are widely documented. These differences indicate that science and religion are two distinct models of knowing and explaining reality. These are the fundamental differences and conflicts that need to be addressed (1995: 541–542).

We agree. The earlier chapters of this book provide a methodology for approaching these different levels and approaches to knowing; naïve acceptance of psychological fashion and dogma has to be addressed as conscientiously as uncritical understanding of particular translations or interpretations of the Bible. Mature Christians should be prepared to face scientific insights about human nature in the light of a reverently critical attitude to Scripture. God's two books, Scripture and Nature, have to be read carefully, both by those concerned with human nature and by those concerned with miracles or creation.

Chapter 12

Our common future

God is creator, redeemer and sustainer of 'all things', and he has given us the responsibility to care for his creation; we are stewards, not passive onlookers. God is separate from the creation. It is wrong to think of the biosphere as an organism in its own right (as assumed by much New Age speculation). Contemporary secular concerns about the environment converge strongly with Christian understandings, but the Christian has a strong underpinning for concern, because of the assurance of being called to look after God's 'good' creation, and because of Christ's reconciling work on the cross, which affects the whole of creation, not only human beings.

Debates about creation and evolution (chapters 6 and 7) have distracted attention from the proper understanding of the environment as God's creation, for whose care we are responsible to God. This has left the way open for a Pandora's box of odd religious ideas, which in turn have raised suspicions about orthodox Christian interpretations of the environment and distracted attention from the obligations of stewardship laid by God on all people (not only those who acknowledge his rule).

Biblical teaching about creation
Ex nihilo

The doctrine that God created 'out of nothing' has been touched upon in chapter 6. Somewhat surprisingly, it emerged as an explicit concept only as late as the end of the second century AD, as a result of debates about Gnosticism. The ancient Jewish tradition was that God created by ordering an existing chaos; that is, God put undefined matter into order. This understanding was followed by the early Christian apologists (such as Justin Martyr, who had an essentially similar interpretation, albeit incorporating Platonic ideas). Tertullian (160–225), however, argued that we have to distinguish three possibilities: that God created the world out of the divine self; or out of something else (which meant that matter existed apart from God); or out of nothing. But if something existed alongside God, this would mean that God would not be absolute; hence God must have created out of nothing.

A few years earlier, Theophilus of Antioch had pointed out that God was far greater than any human craftsman since he did not need material: he just spoke. Irenaeus (130–200) built upon this: God is boundless and gives bounds to all else; he is both transcendent and infinite, and not inherent or contained in the natural world. Nature is demystified and not intrinsically holy.

Nature is distinct from God

Nature is not God (*pantheism*) nor is God contained in nature as well as being beyond and outside it (*panentheism*). Christians have always eschewed pantheism (see Is. 44:14–18; Jn. 1:1–3; Col. 1:16–20), although Teilhard de Chardin seems to have believed in something close to it with his notion of all creation coming together in fulfilment with the glorified Christ at the 'Omega Point'. Suffice it here to say that these ideas are not based on Scripture; they depend on scientific and theological orthogenesis for which there is no evidence, and, although avowedly Christocentric, contain hope but no gospel (D. G. Jones 1969). Panentheism, on the other hand, has a wide following in some theological circles, because it provides a way of envisaging God's immanence without denying his transcendence. Its most influential advocate has been the German theologian Jürgen Moltmann (p. 236). It has been nourished by the process theologians, especially Charles Hartshorne and John Cobb (Birch & Cobb 1981); their ideas have links to liberation theology, with its emphasis on a political eschatology where love and peace come together, and have been particularly influential in World Council of Churches pronouncements (Birch, Eakin & McDaniel 1990). This theology has developed from a conviction that God cannot be impassive to suffering in the world; as he is affected by it, he responds and therefore changes through time. Closer examination, however, shows the difficulty of the concept. As Christian theology it is seriously defective because it relegates Christ's death to that of a mere catalyst within history, and empties it of all eternal significance. Panentheism is attractive to 'greens' because it appears superficially to mesh with scientific thinking, envisaging nature as 'a dynamic process of becoming, always changing and developing, radically temporal in character . . . Anthropocentrism is avoided because humanity is seen as a part of the community of life and similar to other entities' (Barbour 1992: 71).

Nature is not divine

A repeated complaint of environmentalists is that traditional Christian theology has 'de-divinized' the world by separating God from it, and this has legitimized the unsustainable plundering and pollution of creation. In an often quoted and reprinted paper, historian Lynn White argued that 'What people do about their ecology depends on what they think of themselves in relation to things around them. Human ecology is deeply conditioned by beliefs about our nature and destiny – that is, by religion . . . By destroying pagan animism, Christianity made it possible to exploit nature in a mood of indifference to the feelings of natural objects' (1967).

Whatever the historical truth of this statement, it is theologically awry: this world belongs to God by creation (Pss. 8:1–3; 104:5–24) and by Christ's reconciling death (Eph. 1:22; Col. 1:15–20); we are tenants or managers, not owners (Gn. 1:28; Lk. 12:42–48; 20:9–18; *etc*). The command to 'have dominion' was made in the context of men and women 'made in God's image' (Gn. 1:26–27), which must involve a strong element of reliability and responsibility, whatever else it includes.

The non-human world, both living and non-living, is God's *good* creation (Gn. 1:31). Cas Labuschagne identifies as 'one of the most deplorable misconceptions with regard to the biblical doctrine of creation, that creation is usually considered to be anthropocentric . . . [it] is *theocentric*. The creation is God's and the ultimate purpose of creation is not humanity, but rather the embodiment and expression of God's greatness and majesty in the creation' (1991: 122). We must beware of treating the non-human creation as a mere backdrop to God's saving work for sinning humanity.

The converse danger is to sacramentalize nature, as distinct from reifying it (*i.e.* treating it as a 'thing'). Ironically, this can de-divinize nature just as much as reification, since to treat the world as 'God's body' compromises divine integrity and hence denies the integrity of creation and its creatures; it adds nothing to nature's value and lovableness that is not already accomplished through the traditional affirmation of the sacramental presence of the Spirit in the individual. The Bible shows clearly that God loves all creation, human and non-human (*e.g.* Mt. 6:26; 10:29–31; Jn. 3:16), and we are called to do likewise. In other words, we should love nature with *agapē* love, as God loves us, not simply with an *erōs* love that expects a reward and removes the divine.[1]

Nature is intrinsically valuable

Nature has been reconciled to the Father by Christ's work. The danger of nature 'separate from God' is that it becomes treated as a mere 'thing', neutral, or even disordered and in need of quelling. In so far as they had a theology, this was the attitude of the post-Enlightenment Romantics: nature had to be tamed and tidied. This is not the Bible's teaching. Creation is described as 'good' because it is in covenant with the creator (Gn. 9:8–11).

In the Old Testament, God is understood as absolutely transcendent and the world as non-divine. This de-divinization of nature is a basic premise, or prime requisite, for technology and political and social progress. However, de-divinization, as such, tends to lead to de-personalization. This tendency is overcome in the New Testament. In the theologies of John and Paul, God's transcendence is still maintained, although he is now understood in a personalistic way, as personalized in Christ. This personalization of the world in Jesus Christ is particularly marked in the writings of Paul (Eph. 1:9, 10, 22, 23; Col. 1:15–20; Phil. 2:5–11; Rom. 8:18–25) (Faricy 1982: 7; cf. DeWitt 1991).

This means that nature has *intrinsic* worth for the Christian, and provides a motive for creation care in addition to the pragmatic concerns of the non-Christian. These ideas are developed below (p. 225), in relation to the meaning of Christ's work for 'suffering nature'.

Stewardship

We have been given dominion over the non-human creation. We are called to be managers of God's world, not passive bystanders. Adam and Eve were placed in a garden to tend it; plants and animals are for us to eat (Gn. 9:3) as well as to look after for God. Conservation, not preservation, is the Christian mandate. The traditional word for this activity is stewardship; other names are trusteeship, vice-gerency, management or companionship (expressing natural and human interdependence, without excluding the distinctiveness of each). In a pioneering examination of historical attitudes to nature in the western moral traditions, John Passmore (1974) concluded that the belief that we have dominion over nature is a sufficient ethic to cope with the problems of pollution, resources, population growth and conservation, so long as it is interpreted in a suitable humble way. This has been criticized by Robin Attfield (1983) on the ground that this is a restricted and imperfect interpretation of Genesis and implies that everything is made for humankind, nothing else is of any intrinsic value or moral importance, and people may treat nature in any way they like with minimal constraint. 'Instead we should accept that natural processes are not devised or guaranteed to serve humanity, and that manipulating them requires skill and care. Science, technology and cost/benefit analysis should not be abandoned, but we should bear in mind the consequences of our actions' (1983: 5).

The charismatic former Dominican priest, Matthew Fox, rejects 'the stewardship model (that God is an absentee landlord and we humans are serfs, running the garden for God); it does not appeal to the young or to our hearts – it is just one more duty, one more commandment to follow . . . We need mysticism – God *is* the garden.'[2] This is a heterodox and unscriptural rejection of stewardship. A more cogent criticism is that of Mary Jegen:

> Stewardship has failed where it has been reduced to a reasonable way of managing time, talent, and treasure for the sake of the kingdom as we understand it; where it has not created a moral and religious imperative for rectifying the massive structural injustices that make life short and cruel for millions; where it has not moved people to commit themselves to changing the structures that support injustice (1987: 102; *cf.* Hall 1986).

A common claim that can conveniently be included here is that nature has 'rights', and should therefore be treated in the same way as, say, a human person. Although this notion is connected with the idea of the 'worth' of creation, it is difficult to sustain as an absolute principle, because there must be a correlative obligation to maintain the validity of any right. Put in this way, a right in nature is only another way of asserting responsible stewardship to an entity worth respect or value. Although frequently used, rights language tends to confuse the underlying issues, and prevents discussion because rights are usually asserted as somehow inalienable rather than as the consequence of a relationship between right-giver and right-claimer (Taylor 1986; Milne 1986). A right is merely a status, and rights language implies a static relationship between rights-giver and rights-claimer. In fact the relationship between humankind and nature is a dynamic one; it involves a moral involvement on our part, not an unthinking acknowledgment of a situation. To assert that nature has 'rights' is a frequent debating-point, but is unsustainable in practice.

Criticisms of Christian environmentalism

A Christian doctrine of the environment faces two problems: mis-interpretation of Bible teaching, and religious environmentalism dressed up as Christianity.

In the paper already quoted, White declared that 'Christianity . . . insisted that it is God's will that man exploit nature for his proper ends . . . Christianity bears a huge burden of guilt' (1967). White's thesis was based on the premise that our increasing ability to control and harness natural forces was flawed by the assumption that 'we are superior to nature, contemptuous of it, willing to use it for our slightest whim . . . We shall continue to have a worsening ecological crisis until we reject the Christian axiom that nature has no reason for existence but to serve man . . . Both our present science and our present technology are so tinctured with orthodox Christian arrogance towards nature that no solution for our ecologic crisis can be expected from them alone' (1967). But, and this is a key inference, 'since the roots of our trouble are so largely religious, the remedy must be essentially religious, whether we call it that or not'. White went on to conclude that our main hope should be a refocused Christianity, not a wholesale repudiation of it; he suggested that we should return to the 'alternative Christian view of nature and man's relation to it', exemplified by Francis of Assisi's respect for the living world. He proposed Francis as a patron saint for ecologists; in 1980, Pope John Paul II accepted the idea.

There have been even more outspoken condemnations, however. The most commonly quoted one is that of Ian McHarg:

> If one seeks licence for those who would increase radioactivity, create canals and harbours with atomic bombs, employ poisons without constraint, or give consent to the bulldozer mentality, there could be no better injunction than the text ['God blessed them (the newly formed human beings) and said to them, "Be fruitful and increase in number; fill the earth and subdue it. Rule over the fish of the sea and the birds of the air and over every living creature that moves on the ground"' (Gn. 1:28)] . . . Dominion and subjugation must be expunged as the biblical injunction of man's relation to nature (1969: 26).

Max Nicholson puts it similarly: 'The first step must be plainly to reject and to scrub out the complacent image of Man the Conqueror of Nature, and of Man Licensed by God to conduct himself as the earth's worst pest' (1970: 264).

These are gross exaggerations. While there are some Christians who have behaved insensitively towards the environment, a strong stewardship emphasis has run parallel to the 'dominance' theme for much of Christian history. It was implicit in the Celtic church of the early Christian centuries, and is explicit in the Benedictine Rule, which was a major influence shaping society in the Middle Ages (De Waal 1991). It is doctrinally more correct than unfettered human dominance on two grounds: (1) God's command in Genesis was in the context of human beings created 'in his image', which involves trustworthiness and responsibility; and (2) Hebrew kingship was meant to be a servant-kingship, exemplified by the instructions given to David and Solomon, and ideally shown by Jesus Christ; it was not a despotic potency. This is not to say that the attitudes condemned by McHarg and Nicholson have been uncommon. To some extent they can be attributed to rationalization by farmers of their increasing success over 'nature' as technology developed. But the fact that a biblical text was frequently misinterpreted should not be allowed to usurp its correct interpretation. After all, the words of the psalmist that 'the world is firmly established; it cannot be moved' (Ps. 96:10) were taken for many centuries to confirm that the heavens went round the Earth. When it was realized that the Earth went round the Sun, it became clear that the psalmist was talking about the character of God, not basic astronomy (p. 12).

As observers of the world, we have to insist that, although it is fashionable to blame our environmental disasters on Christianity, environmental degradation is almost universal whenever excessive strain is put on natural systems. Leaving aside the horrors produced in Eastern Europe under specifically anti-religious regimes, in other places over-grazing, deforestation and the like on a scale sufficient to destroy civilizations were committed by Egyptians, Assyrians, Romans, North Africans, Persians, Indians, Aztecs and Buddhists. Japan has pollution problems as bad as anywhere in the world. Former EU Commission President Jacques Delors has commented, 'I have to say that the Oriental religions have failed to prevent to any marked degree the appropriation of the natural environment . . . Despite different traditions, the right to use or exploit nature seems to have found in industrial countries the same economic justification' (1990: 22).

Green religion

There has been a trend in recent years to develop various forms of eco-religion, sometimes based on established faiths, but more often on an eccentric ragbag of beliefs. The problems of uncontrolled eclecticism is illustrated by the fate of the Assisi Declarations, produced by some of the major world faiths (Buddhism, Christianity, Hinduism, Islam, Judaism and Baha'i) at the twenty-fifth anniversary celebration of the Worldwide Fund for Nature in 1986. These innocuous and laudable statements led to the establishment of an international 'Network of Conservation and Religion', a useful initiative. But attempts to further the aims of bringing together conservation and religion have led to some highly contentious activities, such as a number of 'cathedral creation celebrations' involving wholly incompatible philosophies, with joint worship by people of different religions, improperly joining different faiths, including monotheists and polytheists. For example, the Coventry celebration in 1988 included a prayer:

226 Science, life and Christian belief

'Our brothers and sisters of the creation, the mighty trees, the broad oceans, the air, the earth, the creatures of creation, forgive us and reconcile us to you.' These phrases are based on Francis of Assisi's Canticle of the Sun; but whereas Francis was concerned to praise the Creator, not to worship nature, they descend to pantheism. Such heterodoxy stimulated a widely circulated 'open letter' signed by over two thousand Church of England clergy, which stated:

> We desire to love and respect people of other faiths. We respect their rights and freedoms. We wholeheartedly support co-operation in appropriate community, social, moral and political issues between Christians and those of other faiths wherever this is possible . . . [but] We are deeply concerned about gatherings for interfaith worship and prayer involving Christian people . . . We believe these events, however motivated, conflict with the Christian duty to proclaim the gospel. They imply that salvation is offered by God not only through Jesus Christ but by other means and thus deny his uniqueness and finality as the only Saviour.

It is important to note the expressed desire to support 'co-operation in appropriate areas'. Creation-care is a duty laid upon all people; it is not an area where Christians should separate themselves from others.

More insidious and difficult to confront are the philosophies underlying the so-called New Age movement. 'New Age' has no precise meaning, but it is claimed to be a sign of the time when the world is moving from Pisces, dominated by Christianity, to Aquarius, symbolizing unity (Groothuis 1986).[3] Characteristic New Age tenets are that (1) *all is one* – invoking subatomic physics as its justification, and ignoring all higher categories of organization; (2) *humanity is God*, which leads to (3) *change in consciousness*, variously called nirvana, satori, self-realization, God-realization, or cosmic consciousness. This means that (4) *all religions are one*, dissolved into a cosmic unity, and implying (5) *cosmic evolutionary optimism*.

The two great New Age anathemas are dualism and reductionism (despite the New Age claim of legitimization from subatomic physics, which is the epitome of reductionist science). Such a faith (if that is an appropriate description) is necessarily pantheistic and relativist (since there are no right/ wrong distinctions); salvation is achieved through self-realization (see pp. 214, 229), so various human potential movements are claimed by New Agers.

The present manifestation of the New Age derives from sundry utopian movements of the eighteenth and nineteenth centuries (especially the Theosophical Society), but it has its immediate roots in the anti-authoritarianism of the 1960s, with its appeals to romanticism as an antidote to the presumed determinism of science. Whereas mainstream thought accepted the need for environmental management and statutory controls, the emerging green movement sought the removal of constraints, allowing life to be lived in harmony with the Earth. Key concepts were balance, stability and peace. A seminal document was E. F. Schumacher's *Small is Beautiful* (1973), with its emphasis on appropriate or intermediate technology. Big business and central government are distrusted. Tradition and authority are suspect, but selectively endorsed in the guise of Earth myths and native customs. Green religion is a passionate animism.

Some of this is healthy. It is right to examine traditions, test authority, and seek to improve the structures of society. But it is too easy to jettison truth in the course of rethinking, and the situation is complicated by the vast spectrum of beliefs and practices between the extreme greens and the most orthodox establishmentarians. Three foci within 'green religion', however, are worth mentioning.

Creation spirituality

Matthew Fox (see p. 223) seeks to unite modern cosmology with 'traditional wisdoms', within which he includes his own background of Dominican mysticism; he frequently quotes medieval visionaries such as Hildegard of Bingen (1098–1179), Meister Eckhart (1260–1329), Julian of Norwich (1342–1415) and Thomas Traherne (1636–74). He argues for the replacement of so-called 'fall/redemption' theology by a creation-centred one, which he sees as an optimistic progression, as opposed to an acceptance of disorder and a need for redemption and reconciliation. For Fox, the biblical God is a sadistic, 'fascist' deity; in his thinking 'we are we and we are God'. Our divinity is awakened through ecstasy – drugs, sex, yoga, ritual drumming or Transcendental Meditation: 'the experience of ecstasy is the experience of God'. Crucifixion and resurrection are transferred from the historical Jesus to Mother Earth; Easter is the life, death and resurrection of Mother Earth, a constantly sacrificed paschal lamb. Fox's religion is one where Christ becomes merely a player on the world's stage; he asserts a form of pantheism where everything is holy and therefore to be worshipped, although he insists that God is more than the Universe and that faith in him is really panentheistic (*i.e.* God is in everything, but is more than everything) (Elsdon 1992; Osborn 1993).

Fox's cosmic Christianity must be distinguished from the more conventional panentheism urged by the so-called process theologians (see p. 220 above). Process theology begins from the premise that God must be open to influence and therefore change by the world he has made; past and present events become joined into a continuum, and redemption becomes part of this process. Consequently, Christ's work is down-played; process theology tends towards a unitarian faith, not a trinitarian one.

Gaia

Green religionists have taken hold of the scientific hypothesis propounded by Lovelock in 1967, that the world and its atmosphere is a self-regulating negative feedback system ('Gaia', after the Greek goddess of the Earth) (Lovelock 1979; 1988). They use it as a justification for the incorporation of human life as merely one element in an interacting but unitary organism. This is not the place to discuss the correctness of the science; Gaia has been an excellent hypothesis in stimulating research to validate or disprove it. The problem has been wild extrapolation from the basic concept, so that the world is seen to be a living creature who can be abused or propitiated. Gaia has become a divine entity to some, worshipped as a goddess, from whose womb we have come. In other words, Gaia science has been used as an intellectual justification for pantheism.

It is not necessary, of course, to endow Gaia with metaphysical properties. Some Christians see the interconnectedness of organic and inorganic systems as an example of the 'anthropic principle', which is that there are too many 'coincidences' in the properties of natural systems for them to have arisen by chance (p. 95). In this sense, the anthropic principle becomes a restatement of the medieval argument from design for the existence of God.

Deep ecology

Some of the more important prophets of green religion are the American founders of the cult of wilderness, notably Henry David Thoreau and John Muir. Muir was born in 1838 into a strict Christian home in Dunbar, Scotland, but was taken at the age of eleven to Wisconsin, whence he went on to be one of the founders of the American National Park system (Austin 1987; Bratton 1993). He was fond of religious language. For example (having just lost a battle to preserve a tract of wild land) he wrote, 'These temple destroyers, devotees of ravaging commercialism, seem to have a perfect contempt for Nature, and instead of lifting their eyes to the God of the mountains, lift them to the Almighty Dollar.' The mantle of these early prophets passed to Aldo Leopold, who turned the notion of respect for nature into a 'land ethic', complementing the ethics of relationships between individuals and with society. Leopold's ideas have in turn been developed by a number of contemporary philosophers, notably the Norwegian Arne Naess (1989) and the American Holmes Rolston III (1988; 1994). Naess contrasts what he calls shallow ecology (which to him merely deals with symptoms, fighting pollution and resource depletion) with deep ecology, based on 'biospheric egalitarianism' (meaning that all things have an equal right to life, although Naess allows self-defence against organisms threatening health or survival). For Naess, deep ecology should explicitly challenge and confront the superficialities of conventional scientific (shallow) ecology; he converges on New Age attitudes by seeing 'self-realization' as a core for fully understanding deep concepts. He believes that deep ecology can be the basis for a comprehensive worldview, linking 'people who ask "ecological questions" in Christianity, Taoism, Buddhism and Native American Rituals'.

Green science

Green science (or science as seen by 'greens') claims to take science seriously, but undoubtedly overemphasizes the interconnectedness of natural systems. Much is made of the concept of a healthy 'ecosystem' which has greater value than its constituent parts; appeal is frequently made to the 'balance of nature'. Herein lies the appeal of the Gaia hypothesis, because its underlying premise is of a massively interacting machine.[4]

There is almost certainly no such thing, however, as a 'balance of nature'. Historically, the idea is based on three concepts (Egerton 1973), all of which are untrue or unhelpful: (1) a parallel between the microcosm of the body and the macrocosm of the living world; (2) the existence of a 'great chain of being' (or 'web of life'), linking all organisms together (this is not the same as the genetic hierarchies which arise through evolution); and (3) a divinely ordained balance, derived from Stoic ideas of the Creator's wisdom and benevolence.

In the early twentieth century, the notion of balance received apparent support from the observations on the succession of (mainly) plant communities, leading to a 'climax'. But it is now clear that there is no such absolute as a climax community: the climax at any time and place is a dynamic relationship with the present environment and past history of that community (McIntosh 1985). This is particularly clear from the plants and animals that live on oceanic islands. Every isolated island lacks some of the species present on continental areas, but the island ecosystems are entirely healthy. Furthermore, experimental disturbance of assumedly species-saturated habitats like tropical forests or coral reefs show that even they are not in some ideal biological equilibrium; the whole is a network of local compromises between death and birth, extinction and colonization, success and failure. One of the founders of modern ecology, Charles Elton, wrote: 'The balance of nature does not exist, and perhaps never has existed. The numbers of wild animals are constantly varying to a greater or less extent, and the variations are usually irregular in period and always irregular in amplitude' (1930: 17). Refinements like chaos theory do not change this picture.

A rather unexpected corollary of this is that reasons to protect the 'biodiversity of nature' are hard to find on purely scientific grounds. Biodiversity is easy to justify on utilitarian assumptions (we may 'need' a species assemblage) or religious premises (we have a responsibility to care for God's world, or, less strongly, for the 'natural world'). It is surprisingly difficult to argue scientifically for the maintenance of the *status quo*; the most convincing arguments are based on the importance of variation for adaptation to changing conditions, but the case for this depends on the factors which maintain variation in populations and communities (Sandlund, Hindar & Brown 1992).

This conclusion does not give us untrammelled licence to disrupt nature and destroy species, because others besides the contemporary human population have interests in the fate of animal and plant communities. What it does is underline the need to go beyond scientific explanation, an emphasis which constantly needs recalling.

Convergence of religion and secularism on the environment

Neither science nor religion by itself can produce the answer to our environmental problems. The toothlessness of science alone was recognized by the lack of impact of the World Conservation Strategy in 1980, which fell into the Enlightenment fallacy that knowledge automatically produces response;[5] it was underlined by the calling forth of the Assisi Declarations (1986) by the Worldwide Fund for Nature and its support for a Conservation and Religion Network; it was made explicit by the Duke of Edinburgh when setting up a consultation on Christianity and the environment, posing the question, 'There must be a moral as well as a practical argument for environmental conservation. What is it?' (Duke of Edinburgh & Mann 1989). The confusions of religion are illustrated by uncertainties about whether to preserve or manage, by the role of established faiths or traditions, and by the selective misuse of scientific data.

Karl Popper has written: 'The fact that science cannot make any pronouncement about ethical principles has been misinterpreted as indicating that there are no such principles, while in fact the search for truth presupposes ethics' (Popper 1978). Is it possible to produce a generally acceptable environmental ethic? The answer to this must be yes. In 1989, the Economic Summit Nations (the G7) called a conference on environmental ethics in Brussels. In the words of its final communiqué, the participants 'benefited from a high degree of convergence between people of different cultures, East and West, and a wide variety of disciplines'. There was absolute unanimity among those present that the main need for individuals and nations alike was to practise responsible stewardship (Bourdeau, Fasella & Teller 1990). This led to a Code of Environmental Practice, based on a simple ethic: *stewardship of the living and non-living systems of the earth in order to maintain their sustainability for present and future, allowing development with forbearance and fairness* (in Berry 1993b: 253–262). In itself, this is an innocuous statement, indeed almost vacuous. However, it entails characteristics common to all good citizens, as well as states and corporations, involving responsibility, freedom, justice, truthfulness, sensitivity, awareness and integrity. In turn these lead to a series of obligations which are its teeth and may involve real cost. The Code is a secular document, produced by a secular group for a secular organization. It was one of the documents submitted as a source paper for the 'Earth Charter' which was intended to preface the work of the 1992 UN Conference on Environment and Development in Rio (but which succumbed to political expediency, and was replaced by an anodyne 'Rio Declaration'). But it was taken almost in its entirety by a working party of the General Synod of the Church of England charged with preparing 'a statement of Christian Stewardship in relation to the whole of creation to challenge government, Church and people'. The General Synod paper began with a statement of Christian understanding:

> We all share and depend on the same world, with its finite and often non-renewable resources. Christians believe that this world belongs to God by creation, redemption and sustenance, and that he has entrusted it to humankind, made in his image and responsible to him; we are in the position of stewards, tenants, curators, trustees or guardians, whether or not we acknowledge this responsibility . . . Stewardship implies caring management, not selfish exploitation; it involves a concern for both present and future as well as self, and a recognition that the world we manage has an interest in its own survival and wellbeing independent of its value to us (*Christians and the Environment*, 1991).

It then drew out the implications of such stewardship in the same way (and in almost the same language) as the Brussels Code. Christian doctrine provides an additional theoretical underpinning for the secular conclusions, but the practical outworking of both sacred and secular is identical – as indeed Christians ought to expect, since they believe that God created, ordained and sustains the world for righteous and unrighteous alike. Orthodox Christian doctrine is that God is both transcendent and immanent: outside and controlling the world, and inside and influencing it (as anyone who prays in faith implicitly accepts). Jonathan Porritt has claimed that the Christian error is to believe in a God far way and remote, whereas the discovery of green religionists is that God is within and intimate (Porritt & Winner 1988: 242–243). Porritt's version demonstrates only too clearly the church's failure to claim and expound sound doctrine, as well as the greens' acceptance of a half-truth as potentially distorting as was the opposite half-truth, exemplified two centuries ago by Paley's 'divine watchmaker'.

Transcendence and immanence

The discussions of 'green religion' and 'green science' are a digression from the Christian understanding of creation, but they illustrate the importance of insisting on the separation of God and creation. As we have seen, the clear teaching of the Bible is that the link between Creator and created is the word of God; creation is not divine, it is not God, and it is now related to God through us ('made in God's image'). The problem ought not to be walking a tightrope between immanence and transcendence, but an unapologetic trinitarianism; the world has been rescued from being merely an object by Christ's work, and is upheld and ordered by God's Spirit. If we see the way forward as a balance between a distant God of absolute power and a confusing pan(en)theism, we will find ourselves repeatedly having to readjust the balance. If, on the other hand, we follow Irenaeus and Tertullian in insisting on a God who alone is self-existent and who created out of nothing, we avoid the dangers of both dualism and a self-centred religion knowable only through self-realization. The contemporary New Age is really a rerun of the Gnostic debate of the early centuries AD.

The recovery of the classical, biblical-theist understanding of God was an important lesson of the evolution–creation debates. It is important to hold on to this in the face of sub-Christian assaults by some 'green theologians'. The most developed form of these is the claim that the Earth (and its biosphere) is a single interacting negative feedback system, personalized as 'Gaia'. But, as we have seen, it is important to distinguish between James Lovelock's Gaia hypothesis in the strict sense of a testable notion about feedback mechanisms, and the extravagant superstructure built upon it which says that the world is a living being (the goddess Gaia), with humans merely one of the elements that influence and interact with it. There is no evidence whatsoever for the latter, and even if Lovelock's ideas were proved correct, this would not diminish the person and greatness of God or our responsibility as his managers. Furthermore, a Gaian god, whether involved in or distinct from the Gaian apparatus, is a unitarian god; there is no room for the redeeming work of Christ or the transforming work of the Spirit.

A lapse into unitarianism, however, is not limited to out-and-out greens. Classical deists used virtually to ignore God's work in redemption and sustaining; there is now a tendency even among apparently orthodox thinkers to assume a degree a interconnectedness in the natural world, with a consequent overemphasis on immanence. From the scientific point of view it needs stating that claims about integrated ecosystem function and long causal chains are widely exaggerated, owing more to Platonic inferences than to modern ecology. We must not allow our scientific beliefs to move us from the biblical position of a trinitarian God who is both immanent and transcendent.

This is crucial when we come back to the basics of Christian environmental concern: there is more to a Christian understanding of the environment than calculating stewardship. Matthew Fox is right: if we are not careful, stewardship becomes just one more command to obey. Indeed, in the industrial world, environmental care is commonly reduced to conformity in meeting statutory requirements, rather than an attitude of respect and moral responsibility. Chris Patten, when Secretary of State for the Environment, described the ideal well: 'The relationship between man and his environment depends, and always will depend, on more than just sound science and sound economics. For individuals, part of the relationship is metaphysical. Those of us with religious convictions can, if we are lucky, experience the beauties as well as the utilities of the world as direct manifestations of the love and creative power of God.'[6]

Can we identify the constituents of this metaphysical relationship? A major part is, of course, experiential. It was awe and wonder that drove Arne Naess to his 'deep ecology'; it was respect for the glories of our world which led such different characters as John Muir, Julian Huxley and Teilhard de Chardin to seek a rationalization for their experiences. It is more than a quest or challenge or a desire for like-companionship that produces escape to the wilds. But we would urge that there is something deeper, towards which wilderness-seekers are groping. Whether the symptoms are middle-class involvement in recycling, countryside protection or ecoconsumerism, or more radical New Age commitments to self-discovery, there is a widespread recognition of a missing 'order' in modern society. In primitive societies, the constant battle to survive means that this 'disorder' is submerged. This may be the reason for 'return to nature' cults; native societies are perceived to have a wisdom and peace that have disappeared from more advanced cultures. It is an illusion, well illustrated by Thor Heyerdahl, who, evacuated after a year on an 'unspoilt' Pacific island in the Marquesa group where he and his wife had found disease, distrust and misery, wrote: 'There is no Paradise to be found on earth today. There are people living in great cities who are far happier than the majority of those in the South Seas. Happiness comes from within, we realize that now. It is in his mind and way of life that man may find his Paradise – the ability to perceive the true values of life, which are far removed from property and riches, or from power and renown' (quoted in Jacoby 1968: 69).[7]

Robin Grove-White, former Director of the Council for the Protection of Rural England, has come to the same conclusion: 'Rather than the environmental agenda being presented to us from on high by science, the

actual selection of issues . . . arises from human beings responding gropingly
to a sense of the ways in which their moral, social and physical identities are
being threatened' (1992: 24). Grove-White identifies the way forward as new
theological understandings of the human person and its needs. Lynn White
said much the same in his lecture over three decades ago: 'What we do about
nature depends on our idea of the man–nature relationship.' But we do not
need *new* understandings; our starting-point is the ancient, universally
established, and often disguised selfishness and pride of the individual. Our
greed is at the root of all environmental damage – sometimes expressed as
personal wants, sometimes through corporate action, sometimes as a simple
desire to demonstrate power. This is common ground to all major religions.
The distinguishing trait of the Christian faith is that God has taken action to
deal with the problem (see *e.g.* Col. 1:16–20). Christians have a particular
responsibility to the environment because of their acknowledgment and
worship of God as creator, redeemer and sustainer. For them, abuse of the
natural world is disobedience to God, not merely an error of judgment. This
means that Christians must examine their lifestyle and work out their
attitudes to the natural world as part of their service and stewardship. It also
means affirming a God who is neither remote nor powerless. The Church of
England Doctrine Commission put it thus:

> To accept God as the Creator of all things implies that man's own
> creative activity should be in co-operation with the purposes of the
> Creator who has made all things good. To accept man's sinfulness is to
> recognize the limitation of human goals and the uncertainty of human
> achievement. To accept God as Saviour is to work out our own
> salvation in union with him, and so to do our part in restoring and
> recreating what by our folly and frailty we have defaced or destroyed,
> and in helping to come to birth those good possibilities that have not
> yet been realized . . . To hold that God has created the world for a
> purpose gives man a worthy goal in life and a hope to lift up his heart
> and to strengthen his efforts. To believe that man's true citizenship
> is in heaven and that his true identity lies beyond space and time
> enables him both to be involved in this world and yet to have a
> measure of detachment from it that permits radical changes such as
> would be scarcely possible if all his hopes were centred on this
> world. To believe that all things will be restored and nothing wasted
> gives added meaning to all man's efforts and strivings. Only by the
> inspiration of such a vision is society likely to be able to re-order
> this world and to find the symbols to interpret man's place within it
> (Montefiore 1975: 77–78).

A fallen world

A legitimate objection to the argument about stewardship and responsibility developed in this chapter is that we live in a fallen world which affects creation as well as our attitude towards it. This should not, however, affect our concern or our involvement.

One of the most important New Testament passages about the relationship between Creator and creation is Romans 8:18–25. This is often interpreted to indicate that the creation is 'groaning' as God's punishment to Adam and Eve, and that the effects of the fall included the introduction of a range of biological and geological 'curses', such as weeds, pathogens and earthquakes. Unfortunately for such exegesis, the Bible is singularly inexplicit about the effects of the fall, with the key exception that death entered the world through Adam's disobedience (Rom. 5:12; 1 Cor. 15:21–22). But, as we have seen, death in the Bible is not primarily about disease and decay; principally it is about separation from God, and only secondarily about physical death. After all, God ordained plants to die to provide food for animals (Gn. 1:29; 9:3), and plant death is biological death just as much as animal death. Paul's point in Romans 8 is that as long as we refuse (or fail) to play the part assigned to us by God (that is, to act as his stewards or vice-gerents here on earth), so long is the entire world of nature frustrated and dislocated; an untended garden is one which is overrun by thorns and thistles.

Charles Cranfield expresses this powerfully in his exegesis of the Romans passage.

> What sense can there be in saying that the sub-human creation – the Jungfrau, for example, or the Matterhorn, or the planet Venus – suffers frustration by being prevented from properly fulfilling the purpose of its existence? The answer must surely be that the whole magnificent theatre of the universe, together with all its splendid properties and all the varied chorus of sub-human life, created for God's glory, is cheated of its true fulfilment so long as man, the chief actor in the great drama of God's praise, *fails to contribute his rational part.* The Jungfrau and the Matterhorn and the planet Venus and all living things too, man alone excepted, do indeed glorify God in their own ways; but, since their praise is destined to be not a collection of independent offerings but part of a magnificent whole, the united praise of the whole creation, they are prevented from being fully that which they were created to be, *so long as man's part is missing,* just as all the other players in a concerto would be frustrated of their purpose if the soloist were to fail to play his part (1974, our italics).

An interesting complement to this is given by Robin Grove-White. Building on the conclusion by Lynn White that 'what we do about ecology depends on our ideas of the man–nature relationship', Grove-White writes that it 'has been largely unrecognized in recent theological discussion of the environmental crisis that the orthodox description of the phenomenon embodies a seriously inadequate conception of human nature at its very centre . . .' (1992: 24, 28). The implication is clear. Without in any way being complacent about the present situation, a full Christian approach to the environment does not need a revamping of traditional doctrines; it should be based firmly and squarely on our divine mandate to be stewards, responsible to God, who is creator, redeemer and sustainer.

Weaknesses in current perceptions
Accountable action

The doctrine of responsibility for our actions is well established in Christian, particularly Protestant, thought. The extent and burden of this responsibility are less firmly embedded; too often our responsibility is seen to be solely spiritual, extending to personal behaviour but not to the care of God's creation. Still less is our role as stewards seen as one for which we shall be judged (Lk. 12:46; 19:24; 20:16). Jürgen Moltmann (1985) has written extensively and helpfully about creation, but is weak on how he regards God's acts in the world and about judgment on earth (Walsh 1987; Deane-Drummond 1992). We can learn much from many passages in the Old Testament which provide an extended parable of the Israelites' misuse of the promised land, so that it never became the intended haven flowing with milk and honey.

Land and wilderness

The Old Testament teaches a strong land ethic. In modern times, a land ethic is particularly associated with the writings of Aldo Leopold. He wrote that we must 'quit thinking about decent land use as solely an economic problem. Examine each question in terms of what is ethically and aesthetically right, as well as what is economically expedient. A thing is right when it tends to preserve the integrity, stability and beauty of the biotic community. It is wrong when it tends otherwise' (1949: 224–225). Unfortunately, many green religions have interpreted a 'land ethic' as a call to preservation. John Muir has become a green prophet – but not a Christian one: he specifically claimed to have turned to Buddhism (Austin 1987: 3). We need to develop a critique of wilderness: is God particularly present there? Is God's ideal for this world a return to a pre-fall paradise?

Conservation and rhythm

The Genesis creation narratives are placed in a seven-day framework; Leviticus 25 extends the Sabbath context, setting out two particular land laws – of Sabbath rest (verses 1–7, 18–22) and of Jubilee (verses 8–17, 23–24). 'The conservation of natural and other resources which is prescribed by this legislation forms the basis of good agricultural and ecological practice.'

This stimulates two thoughts. First, Don Cupitt has commented that 'Religion was more badly shaken when the universe went historical in the nineteenth century than it had been when the universe went mechanical in the seventeenth century' (1984: 58). Time is, of course, the Achilles' heel of natural theology (Mayr 1982: 349); it would be possible for a creator to design a perfect organism in a static world of short duration, but not one which remained perfectly adapted in a world where the environment changes, sometimes drastically. An engine is simply a machine; life has history, and this affected perceptions much more than learning about mechanisms, which were anyway assumed to have been made by God.

Secondly, by creating days, God initiated the fundamental rhythm of the life of humankind. Dietrich Bonhoeffer called the seven-day pattern 'the natural dialectic of creation', and warned how technology is undermining this divine pattern (cited in Blocher 1984: 58).

Doctrinal foundations

There is a convergence between Christian and secular attitudes to environmental attitudes and care. A sensible secularist recognizes the value of the environment to self, community and posterity. Indeed, the 1990 UK Government White Paper *This Common Inheritance* began (most unusually for an official document) with an ethical statement: 'We have a moral duty to look after our planet and hand it on in good order to future generations . . . We must not sacrifice our future well-being for short-term gains, nor pile up environmental debts which will burden our children.' Any non-religious approach, however, can justify the value of nature itself only either anthropocentrically (the usefulness to us of foods, drugs, natural processes such as pollutant breakdown, and so on) or mystically. All religions have a creation doctrine; if we believe that God reveals himself and his work to us (Acts 14:16–17), we should be explicit in proclaiming it (Jb. 38–42).

Conclusion

It is easy to argue about the relative priorities of homelessness, famine, disease, environmental degradation and so on – never mind evangelism, spirituality and worship. These debates are legitimate, although they are somewhat unreal, because God has purposes, not priorities; our difficulty lies

238 *Science, life and Christian belief*

in discerning his purposes (Dt. 29:29; Is. 55:8). It is important, however, not to be sidetracked from our responsibility to God's creation. We ought, first, to focus attention on a God active in the world as distinct from one wholly above the bright blue sky. Secondly, we must re-establish the idea of God as Lord of both the world and of the church, as author of the two complementary books of Scripture and Nature; we cannot be mature Christians unless we read them both. Finally, we. must emphasize our obligation to manage the world for God – and to realize that we sin if we fail in or ignore this task. It is this last lesson which we may have learnt least of all.

Chapter 13

The implications of science

This final chapter brings together the main threads from the preceding chapters, noting some recurring themes for Christianity and science. Christianity is a reasonable faith, requiring the development of a 'Christian mind'; lack of thought in spiritual matters is often a sign of immaturity. Notwithstanding, science has limits, and faith may go beyond reason without being divorced from it. Christians must develop attitudes and behaviour based on their experience as on well as revelation and doctrine, with these experiences being illuminated by a redeemed perception of the world. Scientists who are Christians are scientists who can be enthusiastically involved in their discipline.

Selective attention and selective perception are universal aspects of human behaviour. If it were not so, we should be overwhelmed by the information impinging upon us at any given time. Such selectivity inevitably applies to any assessment we make about the relation between science and religion. Anyone who doubts this should read the book of interviews by Russell Stannard (1996), with scientists involved in the science-and-religion debate. The outspoken atheist chemist, Peter Atkins, had no doubt:

> I think religion kills. And where it doesn't kill, it stifles. Religion scorns the human intellect by saying that the human brain is simply too puny to understand. In contrast, science enables one to liberate oneself; it

liberates the aspirations of humanity . . . [Science] gives people answers that are much more reliable, much more plausible than the obscure arguments that religion provides . . . Science can show that there is not a purpose in the universe, and is not going to waste its time worrying about it (in Stannard 1996: 167–168).

Atkins has a ready ally in Richard Dawkins. For him, religion is like a computer virus, although, as Stannard pointed out, the argument could equally well work the other way round, with atheism as the virus infecting the normal system, since religion seems to have emerged in almost every tribe in every people that has ever lived. When Stannard remarked to Dawkins, 'You seem to be driven to fight almost as religiously against religious belief as I would be on the other side of the fence', Dawkins agreed: 'One of my reasons is that quite a lot of evil is done in the name of religion . . . I think it stems from the fact that religious belief is held with enormous conviction without backing evidence.' Confronted with the suggestion that the vast majority of religious people fully accept scientific explanations – including Big Bang cosmology and the theory of evolution, never mind Stannard's own personal commitment to engage in propagating the public understanding of science, Dawkins replied, 'But I think that people like you are very unusual as religious believers. There is a minority of people sophisticated enough to understand science, to believe in evolution and so on, and to couple it with their belief in God. But there is, I think, a great divide between the sophisticated theologians on the one hand and the people in the pew on the other' (in Stannard 1996: 164–165).

The views of atheists such as Atkins and Dawkins seem extreme against the thoughtful views of agnostics such as the distinguished mathematician, Sir Herman Bondi, President of the British Humanist Association. Bondi was quite happy to agree that he had 'no doctrinal refusal of the idea that the Universe may have a purpose, and may have been designed. Nobody has found a good reason for saying Yes, but equally nobody has found a good reason for saying No' (in Stannard 1996: 173). For him, 'Humanism is, in one sense, a rule for discussion: you cannot bring an argument to a close by referring to a line in the Bible, or in the Koran, or in the writings of Karl Marx.' When challenged by Stannard on his concern that 'You are against religious people being certain about their understanding of God. How would you react to scientists who speak as though science was their God, and that they were on the road to a complete understanding of everything, total certainty?', Bondi replied, 'You will not be surprised when I say I laugh – if it doesn't raise my blood pressure and make me furious . . .' (in Stannard 1996: 175).

Further presuppositions surfaced in discussions with other scientists. The physicist Paul Davies has written that he sees science as a surer path to God than religion. His justification for this is that

> Science and religion, at least as they are being practised in the major institutions, come at the subject matter from opposite directions. Religion is usually based on some sort of doctrine or an ancient text which is meant to contain revealed truths. It's something you have to accept on faith . . . Science, on the other hand, starts from exactly the opposite point of view. You have to accept, as an act of faith, that there is an existing order in nature that is intelligible to us. That is a huge act of faith. But once you have done that, everything else is tentative or provisional (in Stannard 1996: 178–179).

Like Dawkins, he distinguishes between the ordinary churchgoer and the academic. For example, with regard to the Bible,

> To accept that it has important insights for human beings but is not in any sense *God's* description of the world, then that is fine . . . But, of course, many 'ordinary' believers don't see it that way . . . the real gulf is between theologians and ordinary believers who still cling to the Sunday-school notion of God as a cosmic magician who works miracles from time to time, just like any other force or agency in the universe – a very uninspiring view of the Deity, in my opinion. I think it's up to the theologians to come out of the closet and take their message to the people (in Stannard 1996: 180).

We agree, but add that scientists who are Christians also have to come out of the closet and proclaim that there is no necessary conflict between science and Christian belief.

Clearly, both atheists and humanists have presuppositions about belief – as does the philosopher and the open-minded, enquiring scientist who asks religious questions. Philosopher Roger Trigg, for example, has no doubts about this:

> Many people in the field of religion nowadays believe very strongly that religion is [about] constructing rather than discovering. [But] it is an absolute presupposition of religion that it is talking about an independent God . . . I am always very unhappy about the kind of religious belief that fences itself around so that nothing science ever said could disprove it, be a pointer against it – indeed, even be relevant to it . . . theology – which, after all, was once called the 'Queen of the Sciences' – has one thing in common with science:

its subject matter is governed by the nature of objective reality (in Stannard 1996: 182–184).

The views of Roger Trigg are close to those of theologian and philosopher Keith Ward, who told Russell Stannard:

> I think God is known in very ordinary ways, not necessarily anything extraordinary at all . . . like a general spiritual presence, a feeling that everything around you, your whole environment, is shot through with the presence of a conscious being of some sort . . . [Proof of the existence of God] is very difficult. I have taught philosophy for 35 years and I have never proved anything yet. In fact, I seem to *dis*prove more things as I go along, or at least find good reason to doubt them. So I would not think in terms of proof . . . [but rather see it] like committing yourself to some ordinary human person. You wouldn't talk about 'proof'. You would just say, 'This has become the basis of how I live.' I think it's like that with God. You, as it were, 'bet' your life on there being this reality of beauty and truth which calls you onwards. And that's the way you live. And after a while, you can't see any alternative really. But it's not a strict proof (in Stannard 1996: 93–94).

Presuppositionalism

At this point, we need to take formal account of presuppositions, as they have been dealt with theologically by the Dutch scholar, Abraham Kuyper (1837–1920), and particularly his adherents Herman Dooyeweerd (1894–1977), Cornelius Van Til (1895–1987), and their influential but critical expositor, Francis Schaeffer (1912–84). Their emphasis was on the determining influence of one's presuppositions as they affect virtually all rational argument. Van Til argued that to make sense of *any* reality, it is necessary to assume the reality of a 'self-contained' triune God. This led him to criticize traditional Christian apologetics on the ground that they allow sinners to be the judges of ultimate reality. Dooyeweerd believed similarly that science and philosophy can perform their tasks only if they have a sound Christian foundation. Francis Schaeffer showed the inadequacy of worldviews based on non-theistic premises, and thus restored the confidence of many Christians in orthodox theology. He laid more stress on common grace than his predecessors, but, like them, emphasized the all-pervading effect of the fall.

Kuyper and his followers strongly influenced twentieth-century theology, particularly among evangelicals. They provided a counterweight to the fashionable assumption that morality depends on conscience, not absolute values. But they should not be allowed to obscure the traditional insistence that the world (and reality) can be understood by all (Rom. 1:19–20), that although interpretation can be distorted by presuppositions, the underlying phenomena are real and recordable by any scrupulous observer irrespective of his or her religious beliefs. Calvin commented that those who reject every fruit of pagan learning offend the Holy Spirit who has bestowed such gifts on the heathen (*Institutes* III.ii.15).

The limits of science and the nature of revelation

Our understanding of the Bible has to be continually re-examined in the light of secular knowledge, since this affects our background understanding of the world. Time after time, gaps have opened up between science and faith because of failure to do this. We have noted Pope Urban VIII's insistence on a geocentric Universe because of his interpretation of particular scriptures, the creationists' adherence to particular models of creation, and dualists asserting the separateness of body and soul, all on the basis of internally consistent biblical exegesis which neglected other ways of learning about God's work. We are firmly persuaded that we need to read both books of God. The Bible contains all that is necessary for salvation and is our only source of knowledge about the nature of God. But the God of the Bible is also the creator and sustainer of the Universe; we can learn *how* he acts by studying his works in the world and in people.

Conversely, we need to recognize the limits of science. Nobel Laureate and atheist Peter Medawar has spelt this out:

> That there is indeed a limit upon science is made very likely by the existence of questions that science cannot answer and that no conceivable advances of science would empower it to answer. These are the questions children ask – the ultimate questions of Karl Popper. I have in mind such questions as:
>
> > How did everything begin?
> > What are we all here for?
> > What is the point of living?
>
> Doctrinaire positivism – now something of a period piece – dismissed all such questions as nonquestions or pseudoquestions such as only simpletons ask and charlatans of one kind or another profess to be able to answer. This peremptory dismissal leaves one empty and dissatisfied because the questions make sense to those who ask them, and the answers to those who give them, but whatever else may be in dispute, it would be universally agreed that it is not to science that we should look for answers. There is then a prima-facie case for the existence of a limit to scientific understanding (Medawar 1984: 66).

While there is a danger that science may be misused in an attempt to explain everything, it may also be misapplied by Christians to such an extent that it becomes anti-science. Richard Bube has identified the need for a Christian theology 'in spite of science' (1995). He believes the correct way forward is what he calls 'complementary insights', which is similar to the model we described in chapter 4.

Recognizing that neither science nor faith can give comprehensive answers to all questions, how should we relate the two? Writing in the context of the need to learn about persons from science and Scripture, Stephen Evans has described three ways in which Christians attempt to do this. (1) *Re-interpretation.* There are many who wholly accept the so-called scientific view of the world, and therefore have to reinterpret religious data to make them consistent with currently prevailing science. Evans divides them into those who think that scientific and personal views will inevitably turn out to be compatible, and those who believe that personal (or religious) views can in practice be modified to make them fit with scientific understanding. (2) *Humanization.* This occurs very specifically in discussions of the human person. It includes those who want to re-interpret science (of whom some are opposed in principle to 'positivist' assumptions in science) and those who would like to separate so-called hard science from human science, with the latter susceptible to various understandings. (3) *Perspectivalism.* This is the approach of those, like ourselves, who are jealous to maintain the integrity of the scientific enterprise and determined to resist the extension of the scientific method into a metaphysical position, such as 'scientism' or 'evolutionism'. It includes people like Medawar who regard certain areas of reality as off limits to the scientific investigator (*territorialists*), and others who see the scientific approach as merely one of several ways of perceiving reality (*perspectivalists*) (Evans 1979).

Evans himself favours perspectivalism, since 'such a view allows for a unified view of the person which is congenial to the Bible's emphasis on the unity of the person and the resurrection of the body' (1979). (This was the main theme of Evans's enquiry.) While we agree with him in this, we want to go further, beyond perspectivalism and beyond presuppositionalism; we believe we need an intrinsically dynamic model of the interface of science and faith, which reflects the complementary interaction of science and faith in history, as highlighted by Colin Russell and John Hedley Brooke (see chapters 1 and 2). As is often the case in science where two approaches or sets of data or models of action help towards a full understanding of a situation, so with faith. We saw in chapter 8 that the Bible knows nothing of a purely academic approach to life; the Word of God speaks to and is written on the heart, and God reveals his purposes so that we acknowledge his lordship in the day-to-day outworking of all our activities.

An analogy for accepting and using different streams of information is provided by a model for the way the human visual system works (Milner & Goodale 1995). The common inputs from the outside world are subjected to different transformations and proceed through distinguishable channels which are subsequently recombined in the processing systems of the brain. To operate at optimum efficiency, we have to use all incoming information transformed and filtered in ways which reflect the different functions of the incoming streams.

The Bible should not be interpreted as speaking scientifically. The Bible presents truth, and truth is an essential ingredient in any search for reality. But just as scientist has to interpret all relevant data in seeking to draw a conclusion or test a hypothesis, so Bible truths have to be interpreted and reviewed with other valid evidence. This process is necessary at both the intellectual and the personal level. God's teaching comes through combining Bible understanding with information from other sources: our knowledge of the history of life on Earth should bring together our reading of the Bible with what we learn about biological mechanisms and fossil deposits; our understanding of 'life' should link scriptural statements of life in Christ with molecular and medical discoveries about biological life; our 'soul' has to be understood through searching the Scriptures for relevant revelation, and testing this with data from neurology and psychology. The outcome of this process is not – or should not be – an amalgam or compromise, but complementation, involving the essential components of the different approaches. This produces a stronger framework of belief, not an unstable joining of uncertainties.

This is not to claim that a knowledge of science is essential to know God fully. There is an important sense in which God cannot be known. God is beyond our knowing. The Orthodox tradition, which many in the West are rediscovering, reminds us of the foolishness of trying to comprehend and understand God. God is mystery, and the language of the negative (God is invisible, intangible, and so on) reminds us that it is often easier to say what or who God is not than to say who or what God is. A God whom we claim to know wholly through our own reasoning is the God neither of the Bible nor of the Christian church. Such a 'God' is more likely to be an idol made in our own image. But there is a vital sense in which God *can* be known. God is known as God makes himself known. There is a parallel here with our knowledge of human persons. There are people of whom we say, 'I do not know him'; 'She is very hard to know.' Even in human relationships, there is much we can know about a person, without knowing the person. So it is with our knowledge of God. We can know God in so far as God is willing to make himself known. We can know, not the fullness of God, but 'the things revealed' (Dt. 29:29). As Paul says of himself, so we can say of God: he is 'unknown, and yet well known' (2 Cor. 6:9, RSV).

But how does God make himself known? The Christian answer is that this is accomplished supremely and personally in Jesus Christ. Jesus Christ is the one human person in whom God's mind and spirit are perfectly expressed. Jesus is truly 'the image of the invisible God' (Col. 1:15). He is the one by whom we measure every other way in which God is revealed.

Of course, there are other ways in which something of God can be known in this world, ways which we evaluate in terms of their congruence with God's self-disclosure in Jesus. For example, for many people God is known primarily through experience. He is felt to be near us, his presence filling all things. It may be at the times when we experience deep emotions of love or grief that we sense the presence of God near us or within us. These are experiences which Peter Berger calls 'signals of transcendence'. Some psychologists, of whom perhaps Abraham Maslow is best known, refer to 'peak experiences', which religious people often interpret as signs of God (1968). Some of these fit into the categories William James described in his Gifford Lectures at the turn of the century: 'the feelings, acts and experiences of individual men in their solitude, so far as they apprehend themselves to stand in relation to whatever they may consider the divine' (1908). The researches of the late Professor Sir Alister Hardy at Oxford have shown the surprising frequency with which people admit to being aware of 'a presence or power different from their everyday self' (1975). Clearly, many Christian people would want to say much more about their experiences of God than this, and often want to see them in the corporate context of Christian fellowship much more firmly than do either James or Hardy. But all this reminds us that in the ordinary business of living in God's world, there are sometimes extraordinary experiences which function, for more people than we usually imagine, as 'signals of transcendence'.

Then there are experiences of wonder and the excitement of being alive in a wonderful world; of awe and a sense of God's majesty when we are caught up in creation; of the widespread acceptance of moral values; of the communion of love; and of stillness. In many ways, God's presence and his nature can be discerned, appreciated, felt. The proper test by which we measure these experiences and understandings is their congruence with the way God has made himself personally known in Jesus Christ.[1]

Science is concerned in a particular way with reality. Philosophers (and for that matter scientists) argue for hours about what is reality and objectivity. We are told we can never *know* something absolutely, and quantum physicists insist that any event is modified by our observation of it. Notwithstanding, we should not allow ourselves to be distracted by clever arguments; there actually is a real world 'out there'. We may misapprehend it and misinterpret it, but it is indubitably there with its own properties and limitations. We are not free to ignore this world or to treat its existence as a purely academic exercise. For many centuries there was a firm assumption that the whole world, from the motion of falling bodies to the existence of God, could be understood by pure thought: philosophers and theologians strove to derive proofs of the existence of God; thinkers such as Plato declared that nature worked according to certain principles, and sought to deduce from these principles the phenomena of the whole natural world (see chapter 1). Scholasticism changed into science when facts became important, and the notion that theories could be developed independently of the facts was abandoned. This was the revolution whose beginning is conventionally dated to 1543, when the very different works of Copernicus on astronomy and Vesalius on anatomy were published, and the attitudes in them began to take hold.

Put crudely, we need to understand and examine the world in which we live, and to ask ourselves whether scientists can tell us all there is to know about it or whether there are questions which science cannot answer. In an excoriating examination of the atheistic reductionism of Peter Atkins, Richard Dawkins and the theoretical astronomer Stephen Hawking, Keith Ward has shown the logical inadequacy of their assumptions. He argues that 'God is not merely an external watchmaker. God is the sustainer of a network of dynamic interrelated energies, and might well be seen as the ultimate environing non-material field which draws from material natures a range of the potentialities which lie implicit within them' (1996: 57). We agree; we have no doubt whatsoever that only doctrinaire reductionists can deny the existence of key questions about life and its meaning, and we are convinced that answers to these questions will become apparent in time. Notwithstanding, the answers will be more difficult to obtain unless we examine all the evidence. Put another way, we can get a complete picture only when we read both books of God. To quote Trigg again: when he was asked how he saw objective reality in both the religious and the scientific spheres, he replied,

> I personally think the scientists are in the business of discovering the world, not constructing it . . . science and religion are each dealing with the same reality. We all live in one world. Science is trying to discover it, but it is a world which, if religion is right, was created by God. Therefore, I would expect science and religion (although they have different agendas) nevertheless to be saying things that may be relevant to each other (in Stannard 1996: 183).

Guidelines for a constructive relation between science and faith
We live in a world which is in principle understandable

This is almost a truism, but it needs emphasizing, because if our world was not orderly and predictable, science would not be possible. The danger, of course, is that we then go on to assume that everything is in principle foreseeable, and that if we once knew the starting conditions, we could know all the future including our own fate. Modern developments of chaos theory should have disabused us of such an extensive determinism, but even if we do live in a strongly deterministic Universe, there is still room for free will (and therefore for responsibility) and for God to carry out his own purposes. We may be part of a massive geobiophysical machine, but that does not excuse a naïve view of causation. Purpose and divine control act at a different but complementary level to physical linkages. An understanding and acceptance of modern science does not – and cannot – prove anything about the existence and activity of God. As has often been said, the belief that miracles do not occur is just as much an act of faith as the belief that they do occur.

God has revealed himself to us

He has revealed himself in his written Word, as well as in the Word made flesh, Jesus Christ. It surely makes sense that if there is a God, he will want to communicate with us, and the obvious and normal way to communicate is through words. Explicitly, we believe that the Bible is the major channel of communication to us from God. This faith does not depend on the mechanism by which God caused his words to be transmitted. Paul speaks of all Scripture (which meant for him the Old Testament) being 'God-breathed' (2 Tim. 3:16). This was not a simple matter of God using automata, but a process which allowed the writers to make use of all their academic abilities (Lk. 1:1–14).

In fact the inspiration and authority of the Bible are secondary to those of Jesus Christ. *If* Jesus was truly God as he claimed (*e.g.* Jn. 14:10–11; *cf.* Heb. 1:3), and *if* his death was a triumph over death (Mk. 15:37–39; 1 Pet. 3:18; *etc.*) by which our separation and alienation from God can be overcome (Acts 4:12; Heb. 10:14), then the records of him achieve a special significance (*e.g.* Jn. 1:1–14), and the Old Testament has to be interpreted in terms of the accounts and implications of Christ's life, death and claims in the New Testament. All discussion about the accuracy or inerrancy of the Bible is subsidiary to this. We have to begin from the historically well-attested accounts of Christ and decide whether he was who he claimed to be, because if he was not, he was a paranoid megalomaniac. It is not a valid option to regard Jesus as merely a great teacher and example: he was either divine or deluded; and if he was divine, his death was an active victory over evil, not a disaster. In traditional theological language, the crucifixion and resurrection provide substitutionary atonement, making it possible for us to be justified by faith (Rom. 3:22–24; 5:1–2).

Such language and ideas are not foreign to a scientist, because they begin with the evidence we have (Jesus Christ and his nature) and then explore the consequences of accepting the evidence. Any honest person should go from this to check the coherence of the Christian gospel to see if it stands up to any tests of consistency and subjective experience that we can make (such as God's transforming grace and answers to prayer).

We live in God's world, and we are his stewards

By faith, the Christian believes that God created the world; by reason, we learn something of the timings and methods through which the world as we know it has come about. In other words, there is a credible scientific account of chemical, geological and biological evolution which complements the religious account of creation. The Bible tells us something of the meaning of the world in which we live (that is, it deals with 'why' questions); science deals with the mechanisms by which evolution occurred, which are not described in the Bible (that is, science answers 'how' questions).

But Christians go further: they believe that this is God's world by creation (Pss. 24:1; 104:5–9), redemption (Jn. 3:16), and because he sustains it (Ps. 104:27–30; Col. 1:17; Heb. 1:13), and that it is intrinsically good (Gn. 1:31; Ps. 19:1; 1 Tim. 4:4). And one stage even further still is that God has entrusted this creation to us, to be his stewards, tenants, curators, trustees or guardians, whether or not we acknowledge this responsibility. Adam and Eve were placed in a garden to tend it; plants and animals are for us to eat (Gn. 9:3) as well as to look after for God. Conservation, not preservation, is the Christian mandate (chapter 12). In exactly the same way, there is no biblical warrant for a 'hands off' policy for the study of human nature (chapter 9).

Our Lord's teaching is full of examples and exhortations to accept responsibility for our actions in all spheres of life. For example, we are told about the wicked tenants who were not content to manage sustainably the vineyard entrusted to them, but took all the renewable resources for themselves, and attempted to expropriate the capital as well (Lk. 20:9–19). Because of their failure of stewardship, they were punished severely by the owner of the vineyard when he returned to find out what was going on. This is primarily a parable of the way the religious leaders would kill God's Son, but it speaks also of the tenants' task of straightforward environmental management; their poor stewardship was firmly and massively judged.

In the gospels of Mark and Luke this parable is preceded and followed by teaching about Christ's own authority; in Matthew's gospel, another parable about obedience in environmental work is put between the questioning about Jesus' authority and the parable of the wicked tenants, and it is followed by the parable of the wedding banquet, which is also about judgment (Mt. 21:23 – 22:21). We have been given a job to do, and it matters how we do it.

Developing a Christian mind

A recurring insistence in previous chapters has been the reasonableness of Christian faith, properly understood; reason is not only compatible with faith, it is necessary for faith to mature and express itself in both individuals and communities. Our Lord set out the 'first and greatest commandment' as 'Love the Lord your God with all your heart and with all your soul and *with all your mind*' (Mt. 22:37–38). This differs from the original form in Deuteronomy 6:5, which refers to heart, soul and *strength*. Perhaps true strength should be seen as stemming from intellectual understanding rather than physical capability. What is certain is that the Bible reiterates the importance of mind in many places: Paul exhorts us to *think* radically ('Do not *conform* any longer to the pattern of this world, but be *transformed* by the renewing of your minds': Rom. 12:2); Peter castigates ignorance (2 Pet. 3:16), as does the writer to the Hebrews (6:1–2). Guidance involves both faith (Ps. 32:8) and reason (Ps. 32:9). 'Alleluia psalms' like 103 and 104 are sandwiches of understanding in a framework of praise. Lack of thought is a sign of immaturity (1 Cor. 14:20). Paul persuaded (2 Cor. 5:11), reasoned and held discussions with his contacts (Acts 18:4).

The biblical emphasis on reason and thought as necessary for the development of a 'Christian mind' is clear. Problems arise where faith and reason fail to interact and strengthen each other: too often Christians are coldly rational and spiritually inert, or excitable and unable to defend their faith or explore its implications. They do not have a developed 'Christian mind', an epithet popularized by Harry Blamires in a book with that title. He condemned British Christians:

> Except over a very narrow field of thinking, chiefly touching questions of strictly personal conduct, we Christians in the modern world accept, for the purpose of mutual activity, a frame of reference constructed by the secular mind and a set of criteria reflecting secular evaluations. There is no Christian mind; there is no shared field of discourse in which we can move at ease as thinking Christians by trodden ways and past established landmarks (1963).

Mark Noll began his widely read indictment of American evangelicalism:

> The scandal of the evangelical mind is that there is not much of an evangelical mind . . . Despite dynamic success at a popular level, modern American evangelicals have failed notably in sustaining serious

intellectual life . . . Most evangelicals acknowledge that in the Scriptures God stands revealed plainly as the author of nature, as the sustainer of human institutions (family, work, and government), as the source of harmony, creativity, and beauty. Yet it has been precisely these Bible-believers *par excellence* who have neglected sober analysis of nature, human society, and the arts (1994: 4).

Scientists who are Christians have a particular responsibility to nurture and expound their faith, both because of the widespread perception that faith and science are mutually incompatible, but also because their contact with and understanding of creation give them a peculiar ability to relate the two books of God's revelation. This responsibility is not simply advocating 'natural theology' (that is, what can be learned about God from a study of nature by itself), which is disputed among the theologians, but a positive approach to God's work in providence (common grace) as well as his creating, redeeming and sustaining work (special grace) which is commonly perceived as the subject matter of religion.

Dynamic conviction, not dogmatic certainty

In every generation, new light bursts forth from the study of God's written Word. We are indebted to the biblical scholars and theologians who help us in this venture. In every generation the Bible stands up to its own claim of being a reliable guide. It is God-breathed and 'is useful for teaching, rebuking, correcting and training in righteousness' (2 Tim: 3:16). As we continue to explore our wider understanding and our changing interpretations of the implications of scientific discoveries, so our understanding and interpretation of the Bible may change in the light of such new knowledge. It may, for example, warn us off what a particular passage cannot mean, and return us with fresh vigour to discover what it does mean as we compare scripture with scripture with an open mind. Science cannot tell us the details about the involvement of God with creation, but if we accept the authority of the Bible, we must marry our scientific understanding of Universe history with a clear acknowledgment of God's originating and controlling role. Chapters 5 and 6 examine this in detail.

The poet John Milton (1608–74) was a committed, Bible-believing Christian, but he interwove a tremendous amount of the scientific speculation of his time into his poetry, and tied himself (and his contemporaries) in knots. For example, in Book VIII of *Paradise Lost*, he described the frenetic attempts of medieval astronomers to rescue Ptolemy's cosmology, as they

> . . . build, unbuild, contrive
> To save appearances, how gird the Sphear

> With Centric and Eccentric scribl'd ore
> Cycle and Epicycle, Orb in Orb.

Such is the way of dispute and polemic, but not of science. Unfortunately it befuddles everybody. Two centuries after Milton, the American geologist Edward Hitchcock wrote: 'The theologians, having so mixed up the ideas of Milton with those derived from inspiration . . . find it difficult to distinguish between them.' Indeed, the eighteenth century and the first half of the nineteenth were dominated by science being used as a crutch to support the Christian faith, rather than as a scalpel to dissect the natural world, following the assertion of John Locke in *The Reasonableness of Christianity* (1695) that 'Revelation is natural reason enlarged by a new set of discoveries communicated by God immediately, which reason vouches the truth of, by God.' In other words, the Bible was seen as little more than a camera, enlarging things we might be able to perceive in other ways.

Such an approach contains the seeds of all kinds of confusion, yet it is still used; Christians and secularists alike confuse reason and revelation. For Christians, too often ideas and attitudes are implicitly added to the Scriptures to 'save the appearance' of an out-dated model. We must be ruthless in recognizing worldviews and interpretations as being no more than models of reality (chapter 4), and be prepared to change them. Both endless speculation and gratuitous adornments of the Bible are condemned by the apostolic writers (2 Tim. 2:14; Rev. 22:18).

Christian behaviour and science

Our ethics depend on our attitudes. We need repeatedly to examine our understanding of the world, since this determines our attitude towards it and its ways. We then need to make sure that the outworking of our attitudes is consistent. We have touched upon ethical responses in a number of places (notably in chapters 9 and 12), and this is not the place to expand in detail, but it is pertinent to ask what difference it makes to a scientist if he or she is a Christian, and conversely, what difference being a Christian makes to a person who is also a scientist.

These two questions were put to sixteen scientists who had been successful in their profession (we were among the sixteen), and their replies published in a book, *Real Science, Real Faith* (Berry 1991). There was a great variety in their testimonies. As with any group, their experiences, indecisions and disappointments ranged widely. But all were explicit that they were where they were because of God's sovereign hand on them. For them, science was a vocation, just as strong and real as the occupations we normally think of as vocations (medicine, nursing, teaching, evangelism, pastoral ministry, and so on). The common factor for all sixteen – perhaps the *only* common factor –

was the certainty that the God of the Bible both cares and acts, and influences events and people for his own divine purposes.

Does a Christian have a specific contribution in science? Are there Christian principles which indicate the areas where a Christian should choose to work? Should we seek to become involved in fields where we can help humankind as a whole, or perhaps one helpful for our country's needs? For example, the development of new sources of controlled energy would be of considerable benefit for everyone, but it could lead to contentious issues such as disposing of radioactive waste, striking a balance between technology and conservation (as in the siting of wind generators), the use of agricultural land for biofuels, and so on. There are two general questions that we have to answer, which apply to all, but particularly to scientists: (1) what kind of tenants are we when looking after this planet of ours while the 'Lord of the vineyard' is away? and (2) how good are we at being our brother's keeper, and alleviating the suffering of needy humankind?

The fear of the Lord is the beginning of wisdom

The wisdom writings of the Old Testament repeatedly taught that 'the fear of the LORD is the beginning of wisdom' (*e.g.* Pr. 9:10). Psalm 111, which contains 'the research scientists' text' ('Great are the works of the LORD; they are pondered ["studied", RSV] by all who delight in them'; verse 2) concludes with this phrase, thus offsetting the ever-present danger of worshipping the creation by pointing us to the Creator. True Christian affirmation is of a God who is infinite and changeless, who is outside time and therefore knows the future as well as the past, and, having given us free will (and, some would say, thereby consciously limited himself), allows us to make or mar our lives. But – and this is a major qualification – he has actively intervened through Christ's coming to earth and dying for us, so that there is a way out of the morass in which we flounder as we exercise our freedom.

Scientists begin with the evidence that is available – in this case Jesus Christ and his nature – and then explore the consequences of accepting the evidence. An open-minded seeker should go from such a beginning to investigate the coherence of the Christian gospel and see if it stands up to any tests of consistency and of experience that we are able to make.

God has a purpose

God has not merely created the world and thrown it into space (such a God is akin to Paley's divine watchmaker), but puts us here to glorify and to enjoy him for ever. To the unbeliever, this is empty arrogance; to the believer it is a purpose for living.

To discover the purpose of life, we need to combine the consistency of the evidence with our personal experience. In this regard it is interesting that when he was President of the Royal Society of London, Nobel Laureate George Porter wrote: 'Most of our anxieties, problems and unhappiness stem from a lack of purpose which was rare a century ago and which can fairly be blamed on the consequences of scientific enquiry . . . There is one great purpose for man and for us today, and that is to try to discover man's purpose by every means in our power. That is the ultimate relevance of science – and not only of science, but of every branch of learning which can improve our understanding. In the words of Tolstoy, 'the highest wisdom has but one science, the science of the whole, the science explaining the Creation and man's place in it'.

The quest continues today; it is difficult but not impossible. We shall succeed in our quest only if we remain comprehensive realists, that is to say, keeping our minds open to all the relevant evidence. This must include, as a starting-point, that we read the books of both Scripture and Nature. Failure to do that will leave us as those who try to find their way by admiring the scenery but ignoring the map.

In some circles, it is fashionable today to look for wisdom from primitive religions and native peoples. One of the dangers in such an approach, if it is an exclusive approach, is that we are prone to accept uncritically everything that we are told; there is no necessary reason why the ways that they have adopted should be any better, purer or more reliable than our own. It is certainly worth peeling away the irrelevant shells in which we hide. We live in God's world; let us study and rejoice in that world: 'The heavens declare the glory of God; the skies proclaim the work of his hands' (Ps. 19:1).

Those of us who live in great cities, and are shielded from the raw environment, do well to remember God's questions to Job:

> Who is this that darkens my counsel
>   with words without knowledge?
> Brace yourself like a man;
>   I will question you,
>   and you shall answer me.
> Where were you when I laid the earth's foundation?
>   Tell me, if you understand!
> Who marked off its dimensions? Surely you know!
>   Who stretched a measuring line across it?
> On what were its footings set,
>   or who laid its cornerstone –
> while the morning stars sang together
>   and all the angels shouted for joy? (Jb. 38:2–6).

God points us to himself. Science points us beyond its limits. Reason can answer only some of our questions. Our need is not more science, better reason or great faith; it is faith in a great God.

Notes

1. Hebrew-Christian and Greek influences on the rise of modern science

1. A tendency on our part to oversimplify complex issues in the interests of brevity can be corrected by referring to one or more of the readily accessible, detailed and authoritative reviews (published during the past thirty years) which trace the rise of modern science. Foremost is Professor R. Hooykaas's *Religion and the Rise of Modern Science* (1971). A more recent account which acknowledges its indebtedness to Hooykaas is Colin Russell's *Cross-Currents: Interactions between Science and Faith* (1985), the first three chapters of which summarize some of Hooykaas's main conclusions. More recently still is John Hedley Brooke's *Science and Religion: Some Historical Perspectives* (1991), in which he argues that recent historical scholarship has revealed such a rich and complex web of links between science and religion that it is often virtually impossible to separate their intertwined strands.

2. Lloyd 1970: 125–146 (especially 131–133 on the non-practical nature of Plato's and Aristotle's approach to science; on experiment and observation see Lloyd 1979; 1991, especially 70–99, 299–332).

3. Diderot (1713–84) commented that a deist is someone who has not lived long enough to become an atheist: perhaps that is why there are more atheists today!

4. N. Carpenter, *Philosophia Libera* (Oxoniae), 1622, *Praef.*, quoted by Hooykaas 1957: 20.

5. Russell suggests that its persistence lies in the ceaseless campaigning of Huxley and his allies, and in two books written in the late nineteenth century: *A History*

of the Conflict between Religion and Science by J. W. Draper (1875) and *A History of the Warfare of Science and Theology in Christendom* by A. D. White (1896). Both books achieved wide circulation, yet Russell asserts: 'Today the historical views of Draper and White are totally unacceptable, not merely because of many factual aberrations, but much more because they represent a long-demolished tradition of positivist, Whiggish historiography.' Both books were written in the USA, and Russell argues that in both instances the conflict and warfare metaphors arose from within each writer's own personal and social circumstances, and hence were largely culturally determined.

6. Scopes had not disputed the facts. While the school principal was ill, Scopes had filled in for him, using George William Hunter's *Civic Biology* – a book adopted for all schools by the State Textbook Commission. Scopes later wrote that he did not remember if evolution had been discussed, but, since biology was inseparable from evolution, he agreed he must have taught evolution. At the urging of local free-thinkers and promoters, he reluctantly agreed to let his name be used to generate a test case of the constitutionality of a new State law banning evolution. He knew this would ignite a controversy.

The trial was a bigger circus than expected. The prosecutor was William Jennings Bryan, three times Democratic presidential nominee and a former Secretary of State. Before the trial he announced that it would determine whether evolution or Christianity survived. The defence lawyer was Clarence Darrow, a leading criminal lawyer, and well known as an agnostic.

In fact, the issue was never joined legally. The judge refused to allow the defence witnesses to testify – theologians, biologists, anthropologists and palaeontologists who had come to Dayton to defend Darwinism. The issue, the judge insisted, was simply whether Scopes had taught evolution. Scopes conceded that much. Notwithstanding, Bryan wanted to make it a trial of Darwin *versus* the Bible, despite the judge's reluctance, and Darrow gleefully agreed. He led Bryan into illogical, untenable corners time and time again, when Bryan insisted on identifying himself as an 'expert' in biblical science. Bryan tried to prove that anyone could interpret Scripture, but he was no match for Darrow; and although Bryan won the case, he was humiliated and mocked in the press around the world. He refused to answer questions about the age of the Earth, the antiquity of well-known archaeological sites, and so on. 'I do not think about things I do not think about,' said Bryan. 'Do you think about things you do think about?' retorted Darrow.

In fact, evolution was victorious if the debate were judged forensically rather than legally, and Bryan emerged a rather tarnished defender of the faith. He died a short time later, an old statesman reduced to a laughing-stock in the press, accused of leading his followers to disaster.

7. *The Genesis Flood* stimulated overt 'creationism'; it apparently showed a 'way round' standard geology. In 1963, ten members broke away from the American Scientific Affiliation and formed the Creation Research Society. This grew rapidly, and within ten years was claiming a membership of 450 voting members (with postgraduate degrees in science) and over 1,600 non-voting members. All

members had to subscribe to an official Statement of Belief which summarized the basic tenets of the movement as a whole:

1. The Bible is the written Word of God, and because it is inspired throughout, all its assertions are historically and scientifically true in all the original autographs. To the student of nature this means that the account of origins in Genesis is a factual presentation of simple historical truths.
2. All basic types of living things, including man, were made by direct creative acts of God during the Creation Week described in Genesis. Whatever biological changes have occurred since Creation Week have accomplished only changes within the original created kinds.
3. The great Flood described in Genesis, commonly referred to as the Noachian Flood, was an historic event worldwide in its extent and effect.
4. We are an organization of Christian men of science who accept Jesus Christ as our Lord and Saviour. The account of the special creation of Adam and Eve as one man and woman and their subsequent fall into sin is the basis for our belief in the necessity of a Saviour for all mankind. Therefore, salvation can come only through accepting Jesus Christ as our Saviour.

In 1970 Christian Heritage College and its research division, now known as the Institute for Creation Research, were established in San Diego, California. The Institute declared that it 'recognizes the Bible as the source of all truth and meaning of life and God as the Creator and Sustainer of all things. Its goals are to re-establish these principles in the educational and scientific worlds.' It was, it claimed, the first known time in history that an educational and research centre had been founded strictly on 'creationist' principles and purposes. Projects have included a search for Noah's Ark on Mount Ararat; research into fossil anomalies; field studies on inverted geological sequences; and library research in current scientific publications dealing with origins.

8. In 1966, membership of the EPM was 200; stimulated by influences from America it rose to around 850 by 1970. It changed its name to the Creation Science Movement in 1980. Following a visit from Henry Morris, a Newton Scientific Association was formed in 1972. Adopting the American practice, its policy was not to include quotations from the Bible in its lectures or literature. A British Biblical Creation Society was formed in 1977 (with a membership of 700 in 1982).
9. Quoted from Nathaniel Carpenter, *Philosophia Libera* (Oxoniae), 1622.

2. *God, creation and the laws of nature*

1. Polkinghorne recounts how the seeming irrationality of the superconductivity state made sense only when it was realized that 'there was a higher rationality than that known in the everyday world of Ohm. After more than fifty years of theoretical effort, an understanding of current flow in metals was found which subsumed both ordinary conduction and superconductivity into a single theory. The different behaviours correspond to different regimes, characterized by different organizations of the states of motion of the electrons in the metal. One regime changes into the other by a phase change (as the physicists call it) at the critical temperature.'

Polkinghorne finds the notion of regimes a useful analogy for the new coherent understanding that comes when events such as miracles are set within their proper context of the basic claims of Christianity: 'Christianity claims that God was in Christ in a unique way. If that is true it is to be expected that unprecedented events might occur, for Jesus represented the presence of a new regime in the world. Along these lines I believe that it is possible to form a coherent picture of God's activity in the world that embraces both the fact that in our experience dead men stay dead and also that God raised Jesus on Easter Day (and, I believe, that the tomb was found empty).'

3. The scientific enterprise

1. Banner quotes Cupitt: 'Most people begin by thinking of their religious beliefs as being "literally" or descriptively true; as describing – however inadequately – real beings, forces and states of affairs. Thus there is supposed to be an objective God, another and higher world, a life after death and so forth. We call this naïve kind of belief theological realism. Natural as it seems, there must be something wrong with it, for nowadays we have highly refined tests and standards for what is to count as knowledge, and by these criteria no religious belief whatever today belongs to the public body of tested knowledge' (Cupitt 1984: 18).

   As Banner says, 'The assumption that there are clear tests for what is to count as knowledge, tests which religious belief fails, sounds like a view which comes from the simple world in which science is capable of proof whereas religious belief is beyond verification' (Banner 1990). Our belief is that if Cupitt and others like him listened more carefully to philosophers of science interested in the logic of scientific discovery and in the status of scientific knowledge, they would have been less inclined to adopt such a simplistic view.

2. Shapere comments: 'In its concentration on technical problems of logic, the logical empiricist tradition has tended to lose close contact with science, and the discussions have often been accused of irrelevancy to real science. Even if this criticism is sometimes overstated, there is surely something to it, for in their involvement with logical details (often without more than cursory discussion of any application to science at all), and in their claim to be talking only about thoroughly developed scientific theories (if there are any such), and in their failure (or refusal) to attend at all to questions about the historical development of actual science, logical empiricists have certainly laid themselves open to the criticism of being, despite their professional empiricism, too rationalistic in failing to keep an attentive eye on the facts which constitute the subject matter of the philosophy of science' (Schapere 1981: 31–32).

3. Banner concludes that logical positivism was one of 'a succession of unsatisfactory philosophies of science which were taken to establish the need for reductionism in theology. Thus logical positivism was, and still is, used to support the popular view that traditional religious belief is epistemologically very different from scientific belief and so should be abandoned in favour of the atrophied theology of a Braithwaite or a Cupitt or a Van Buren' (1990).

4. Ayer continues: 'Popper therefore has no truck with any talk of confirmation. All

the same, he is willing to say that a hypothesis is corroborated it it passes a severe test. We are not, however, supposed to infer that its being corroborated makes it any the more credible . . . But this is very strange. For what would be the point of testing our hypotheses at all if they earned no greater credibility by passing the tests? It is not just a matter of our abiding by the rules of a game. We seek justification for our beliefs, and the whole process of testing would be futile, if it were not thought capable of providing it' (1946: 134). As regards induction, Ayer comments: 'Not only that, but the whole pretence that we do not reason inductively becomes ridiculous when we consider how much inductive theory is built into our ordinary ways of speaking . . . [Although] Popper gives a luminous account of at least one form of scientific procedure, the basis of his system is insecure' (1946: 134).

5. Lakatos described his approach as 'best presented by contrasting it with falsificationism and conventionalism, from both of which it borrows essential elements . . . The methodology of research programmes presents a very different picture of the game of science from the picture of the methodological falsificationist. The best opening gambit is not a falsifiable (and therefore consistent) hypothesis but a research programme. Mere "falsification" (in Popper's sense) must not imply rejection. Mere falsification (that is anomalies) are to be recorded but need not be acted upon. Popper's great negative crucial experiments disappear; "crucial experiments" is an honorific title, which may, of course, be conferred on certain anomalies, but only long after the *events,* only when one programme has been defeated by another one. According to Popper, a crucial experiment is described by an accepted basic statement which is inconsistent with a theory – according to the methodology of scientific research programmes, no accepted basic statement alone entitles the scientist to reject a theory' (1970).

6. Gottfried and Wilson, seeking to be fair to the Edinburgh school, believe that it 'does not see itself as opposing science, or questioning the integrity of scientists. But it contends that scientific knowledge is only a communal belief system with a dubious grip on reality, and it is this claim that we address from our perspective as physicists' (1997: 545). From their perspective as physicists, Gottfried and Wilson argue that 'Pickering's statement about the irrelevance of twentieth-century science stems in part from a recurring misunderstanding in sociological studies – that what scientists see as progress often entails abandoning familiar but perplexing phenomena; in this instance the more commonplace phenomena that dominated the "old" physics were supposedly abandoned in favour of more esoteric phenomena in the stampede to the "new". But they were not abandoned, they were set aside, as has often been fruitful in physics . . . Galileo set aside friction. Bohr's demonstration that the hydrogen spectrum holds the key to atomic physics led to a shift of interest towards spectroscopy, but after the development of quantum mechanics the "abandoned" phenomena (for example in solids) returned to centre stage. The "new" particle physics was, in large part, born of the recognition that the "old" phenomenology was less accessible than that on which the "new" physics concentrated. This venerable strategy is being vindicated now with lattice QCD, a numerical approach that is yielding the first

fundamental understanding of hadronic spectra' (1997: 546). They also point out that whatever quarrel scientists may have with Edinburgh pales in comparison with the hostilities between Edinburgh and some philosophers of science, a conflict that can make a scientist feel like an Algonquin whose hunting grounds are being fought over by two colonial powers' (1997: 547).

7. The analysis here closely follows that given by Van Till (1988: chapter 1; Van Till *et al.* 1990: chapter 5).

4. Explanations, models, images and reality in science and religion

1. 'The nature and content of the knowledge relation will be determined by the two poles between which it extends, by the object pole as well as by the subject pole. The possibility of discovering and describing order, uniformity and constant relations within the phenomena, is certainly due to the nature of the object. But it is equally due to the nature of the human intelligence, which enables men to build up a scheme of logically coherent concepts, which are adequately related to the features of the object. This conception means that the objectivist as well as the subjectivist interpretation of physical knowledge have both to be rejected as being tendencies to ignore, or anyhow to minimise, the significance of one of the two poles between which the knowledge extends' (G. J. Sizoo, conference paper quoted in Jeeves 1969: 59).

2. 'Over and above the recognition of abstract patterns characteristic of the sciences of inanimate matter, the practice of biology demands the recognition of individual living things, and analysis in biology is always analysis *within* the context set by the existence of such individual living things. Thus the recognition of individuals adds to the subject matter of biology a logical level missing in the exact sciences, and at the same time limits the range of analysis to the bounds set by the acknowledgment that individual living things exist. *I do not* mean that at some mysterious point analysis will have to stop, but that an analysis of an organism which analyses the organism *away* would contradict itself by destroying its own subject matter. Nor do I mean that when we recognize individuals we are adding some mysterious vital something that comes from I know not where, but that we *are* affirming the existence of something which is more than a brute fact, in the sense that we acknowledge it as an achievement: as an entity that succeeds or fails relatively to standards which we set for it. It is a good or a bad specimen of something, *Cepaea nemoralis* or *Spiraea van loutiens*. We recognize it as an individual in respect to its trueness to type; no matter how far analysis may proceed, this recognition will always be essential. Otherwise we should not know what we were analysing' (Grene 1966: 207).

3. Grene refers to the increasing relevance of biochemistry to the understanding of genetics, and concludes: 'As genetical research proceeds, along with specification, the nature of the whole, too, makes itself felt. The parts are the conditions for the whole, which certainly could not exist suspended in some heaven of essences without them; but it is the whole that *explains* the part, not the parts the whole. The whole is the system (the organism) that makes the parts what they are, even though the parts are the conditions (in traditional language, the material causes),

for the existence of the whole . . . Biological explanation, however – and that is my point in reference to genetics – entails the recognition not only of systematic connections – between *such* genes and *such* phenotypes – but of individually existent systems: organisms existing as unitary four-dimensional wholes, as individuals with a life history in a particular portion of space time' (1966: 208).

4. In answer to the attack that this approach smacks of 'teleology' and of 'vitalism', Grene points out: 'In understanding and explaining organic phenomena we are proceeding in a fashion different from the way in which we proceed when explaining physical phenomena, [so] this suggests also that *what* we are explaining is in fact existentially and historically different . . . The discontinuity of emergence is not a denial of continuity but its product under certain conditions' (1966: 211–212).

5. The God of the physical Universe

1. The evidence for and against the Big Bang model is reviewed by Wilkinson and Frost (1996).
2. Such an argument against the existence of God is put by Atkins (1994).
3. See Van Till 1986; J. Wright 1994; Houghton 1995; Wilkinson & Frost 1996.

6. Creation

1. Interpretations of the Genesis 'day':

    1. Literally. The traditional interpretation is, of course, that the days represent literal twenty-four-hour periods. Defenders of this view treat the Genesis account as history, arguing that the text contains no indication of figurative language; they maintain that those who take any other view have given in to the spirit and mind of an apostate world. They quote other parts of the Bible as apparently treating the Genesis story in the same way – verses such as Ex. 31:14 (the Sabbath commandment), Mt. 19:4 (Jesus' comment on the relation of man and woman), and 2 Pet. 3:5 (the reminder that the earth was formed out of water). The latter two passages, however, are not really relevant and it is not clear anyway that they imply a literal reading. Even the Exodus passage does not require a literal interpretation of 'day'; it merely refers back to Genesis, and in the repetition of the Sabbath law (Ex. 31:17), clear use of an anthropomorphism is made, when the text says that 'the LORD . . . rested'. The second version of the Ten Commandments (Dt. 5:12–15) replaces the reference to creation with the memory of Israel's slavery in Egypt, which does not contradict Ex. 20, but warns of too close a link between the work of the Creator and the weekly rhythm of human life.

    It is clearly possible to interpret the Gn. 1 'day' as a twenty-four-hour period, but it is hard to maintain that no other interpretation is legitimate, and dubious to reject all the scientific evidence about the Earth's antiquity.

    2. The gap theory. The Scottish divine, Thomas Chalmers, seems to have been the first person to suggest that the six days were days of reconstruction, not of creation. He argued that a catastrophe happened between the creation of the heavens and the earth (Gn. 1:1) and the formless and empty earth described in Gn. 1:2. Indeed, there could have been several catastrophes, perhaps corresponding to the mass extinctions revealed in the fossil record, and therefore (it is argued) reducing any conflict with science about the age of the Earth. This 'gap theory' was incorporated by Scofield into his 'annotated Bible', and achieved wide circulation. It is, however, exegetically unsound. It depends on reading Gn. 1:2 as 'the earth became', which is inadmissible. Also, it requires the verb translated 'make' in Gn. 1 (and Gn. 2:2; Ex. 20:11) to have the meaning 'remake'. There is no justification at all for this.

3. *Days of revelation.* P. J. Wiseman proposed that, since 'the Sabbath was made for man' (Mk. 2:27; clearly the Creator did not need a day's rest), it was intended for human rest. 'Then it is only reasonable to suppose that what was done on the "six days" also had to do with man; and if with man, then obviously on the six days God was not creating the earth and all life, because man was not in the world when these were being created' (1948: 40). Wiseman then went on to suggest that the implication of the repeated phrase 'God said' was that it means that God *told* man what he had done in times past; the six days of Gn. 1 therefore become days of revelation. 'God said', however, is always followed by the *making* of something, not the *revealing* of anything. Other scholars have not followed Wiseman.

4. *The concordist theory.* The suggestion that the days of Genesis represent geological epochs goes back at least to the early centuries AD and is commonly repeated. For example, Payne Smith in Ellicott's conservative evangelical *Commentary* was explicit that 'A creative day is not a period of twenty-four hours, but an *aeon*, or period of indefinite duration, as the Bible itself teaches to us . . . By the common consent of commentators, the seventh day, or day of God's rest, is that age in which we are now living. So in Zechariah 14:7 the whole Gospel dispensation is called "one day"; and constantly in Hebrew, as probably in all languages, *day* is used in a very indefinite manner, as for example in Deuteronomy 9:11' (1897: 13).

While not going along with Ellicott in accepting that the seventh day is necessarily the gospel age, it is certainly true that the Hebrew word *yôm* does not always mean a twenty-four-hour period. In the Old Testament it may be a time of special divine activity ('day of the LORD', Is. 2:12), an indefinite period ('day of temptation', Ps. 95:8), or simply a long period (Ps. 90:4). In Gn. 2:4, the word used for the whole span of time during which God was creating is the same as the one used for a single day in Gn. 1. Probably most apologists would agree that it is not particularly meaningful to speak of a twenty-four-hour day in terms of an eternal God who is outside time.

The difficulty with the concordist theory arises when the details of the fossil and scriptural records are matched with each other. Although there is considerable agreement between the order of events, there are discrepancies. For example, trees (Day 5) and flying organisms (Day 5) precede terrestrial animals (Day 6), contrary to scientific expectation. But the biggest disagreement is the creation of the Sun and stars on the fourth day after that of dry land and its vegetation on Day 3. The usual justification of this is that on Day 4 God dispersed a thick covering of cloud, so that the Sun and Moon and stars were revealed to Earth for the first time. The snag here is, as with the 'gap theory', that advocates of a particular view would like to change the meaning of a simple word; in this case to change 'make' in verse 15 to 'reveal'.

There is no justification for this; there is a perfectly good word in Hebrew for 'appear', which is used in verse 9 (Lucas 1989: 85–86).

5. *The literary interpretation.* The idea that the Genesis 'days' represent a framework for the creation account, rather than an indication of time, sets it apart from the other four interpretations. It was embraced by Augustine, who believed the six days were 'a sextuple confrontation of the angelic nature with the *order* of creation'. It is held by respected modern evangelical commentators, such as J. A. Thompson, who affirms that 'the whole [of Gn. 1] is poetic and does not yield to close scientific correlations' (*NBD*: 2). E. F. Kevan, a former Principal of the London Bible College, in the first edition of *NBC* (1953: 76), says: 'The biblical record of creation is to be regarded as a picturesque narrative, affording a graphic representation of those things which could not be understood if described with the formal precision of science. It is in this pictorial style that the divine wisdom in the inspiration of the writing is so signally exhibited. Only a record presented in this way could have met the needs of all time.'

In the second edition of *NBC* (1954), Meredith G. Kline wrote: 'The prologue's [Gn. 1:1 – 2:3] literary character limits its use for constructing scientific models, for its language is that of simple observation, and a poetic quality, reflected in the strophic structure, permeates the style. Exegesis indicates that the scheme of the creation week itself is a poetic figure and that the several pictures of creation history are set within the six work-day frames not chronologically but topically. In distinguishing simple description and poetic figure from what is definitely conceptual, the only ultimate guide, here as always, is comparison with the rest of Scripture.'

Francis Schaeffer goes even further when he discusses time in the early chapters of Genesis. He points out that time is not used chronologically in these chapters, nor are genealogies complete or even in the expected order. He is emphatic in his conclusion: 'In regard to the use of the Hebrew word *day* in Genesis 1, it is not that we have to accept the concept of the long periods of time that modern science postulates, but rather that there are no clearly defined terms upon which at this time to base a final debate . . . Prior to the time of Abraham, there is no possible way to date the history of what we find in Scripture' (1973: 124).

Henri Blocher has described the literary interpretation and its attractions well: 'The literary interpretation takes the form of the week attributed to the work of creation to be an artistic arrangement, a modest example of the anthropomorphism that is not to be taken literally . . . Two centuries ago Herder recognized the powerful symmetry between the two triads of days: Day 1 corresponds to Day 4, Day 2 to Day 5, Day 3 to Day 6. Corresponding to the light (1) are the luminaries (4); to the creation of the expanse of the sky and the separation of the waters (2) correspond the birds and the fish (5); and to the appearance of the dry land and of vegetation (3) correspond the land animals including mankind together with the gift of food (6). Medieval tradition had recognized the broad pattern, since it distinguished the work of *separation* (Days 1–3) from the work of *adornment* (Days 4–6). It would be better to speak first of *spaces* demarcated by divine acts of separation, then of their corresponding *peopling*. It can also be stressed that only the creatures of the second series are mobile (some speak of the immobile creatures for

Days 1–3 and of mobile creatures for the rest) . . . The theological treasures of the framework of the Genesis days come most clearly to light by means of the "literary" interpretation. The writer has given us a masterly elaboration of a fitting, restrained anthropomorphic vision, in order to convey a whole complex of deeply meditated ideas' (1984: 51).

2. First, why are the Sun and Moon not called by their names, but referred to only as 'lights'? There are perfectly good, common words for 'sun' and 'moon' in Hebrew. The probable answer is that in the Semitic languages the words 'sun' and 'moon' are also the names of gods. The peoples around the Hebrews worshipped the heavenly bodies as gods and goddesses. Gn. 1:14–19 is an attack on all such thinking. The heavenly bodies are simply 'lights' (just like enormous oil lamps) created by the God of Israel to serve humans as calendar markers.

   Secondly, the Hebrew verb *bārā'* (create), which in the Old Testament is used invariably with God as its subject, occurs in verses 1, 21 and 27. It seems understandable that it should be used in the initial statement about God's creative work in verse 1 and of the final act (the creation of humans) in verse 27. But why is it used in verse 21? The most convincing answer has to do with the significance of sea monsters in one of the creation stories of Babylon (and probably Canaan too). Here the creator god has to subdue the forces of chaos, depicted as sea monsters, before creating the heavens and the earth. Gn. 1 rejects this by stressing that the sea monsters are just part of God's creation. He did not have to fight and subdue them; he made them.

3. 'Understanding evolution requires the acquisition of a full-fledged biological education as well as some intensive study in evolutionary thinking. Unfortunately, most of the critics of evolution have only a shallow or narrow knowledge of biology, and their writings on the subject of evolution reflect it. These are often physical scientists, engineers, philosophers, or theologians who are reacting to the philosophical extensions of evolution. Convinced that biological evolution must be false because of some of the claims of philosophical evolutionism, many critics direct their attention to the biological arena, where evolution is so thoroughly embedded that attacking it is like attacking most of biological thought' (R. T. Wright 1989: 118).

4. In 1860, Richard Owen, in his exchanges with Thomas Huxley, and seeking from the highest motives to defend human dignity, argued for humanity's uniqueness on the grounds that 'the great-ape does not possess a hippocampus-minor' (Desmond 1994: 20). His attempt we now see as misplaced. But his questions posed then remain relevant.

   Clearly, we differ from apes much more extensively than might be supposed from the similarity of our genes (p. 207). For example, the size difference between human and chimpanzee brains is greater than that between the brains of the chimpanzee and the shrew. Absolute size, however, is not a good ground from which to try and argue the uniqueness of *Homo sapiens*. Whales have much bigger brains than we have; so have elephants. Nevertheless, the human brain is 3.1 times as big as expected for a non-human primate of the same weight. As we shall see (chapter 11), the distinctiveness of humans is not contested by either Christian or non-Christian scientists. There are different emphases about what features of humanness are particularly significant, but language and the capacity for internal symbolic manipulation come high on most scientists' lists.

5. In his Gifford Lectures, the late Professor W. H. Thorpe, one of the most distinguished animal behaviourists, listed a number of abilities which had confidently been asserted in the past as absent in animals (1974: 271–272). His list included the following: animals could not learn, they could not plan ahead, they could not conceptualize, they could not use tools, they have no language, they cannot count, they lack artistic and all ethical sense. Today, in the light of the evidence gathered by ethologists and psychologists, it would be very difficult to maintain any item in this list. Thorpe concentrated his attention on the six levels of mental activity described by Hobhouse (1913), of which four could be discerned in animals, while the remaining two were regarded as characteristic of humankind. Hobhouse called these two 'the correlation of universals' and 'the correlation of governing principles'. They involve a recognition of abstract moral law or, in Thorpe's words, 'eternal values which are in themselves good . . . of course we can find in the higher social animals, such as wolves, behaviour which appears altruistic, unselfish, indeed "moral". Nevertheless, I believe at this level we can see a difference between the minds of humans and of present-day animals, and that in Hobhouse's last two categories we have reached a distinction which we can for the time being at least regard as fundamental' (Thorpe 1961: 44–45). See also below, pp. 204ff.

6. The creation of woman is described in Gn. 2:21. Eve was formed from Adam's side. John Rendle-Short says: 'This must be historical truth or myth . . . It is naïve to dismiss the story of Eve as myth and assume no violence has been done to Scripture. At least seven doctrines of fundamental importance to the whole human race, but especially to Christians, are directly founded on the fact that Eve was created out of Adam' (1981: 37).

This comment betrays a failure by Rendle-Short to recognize that God may communicate by means other than historical events, just as that Jesus taught by parables as well as by miracles. If God is truly omnipotent, clearly he could have made Eve from one of Adam's ribs, but that does not mean that he actually did so. Modern man has the same number of ribs as modern woman. The emphasis of the Genesis account is on the similarity of the man and the woman, on their close kinship, and on their possession of an identical essence. And as Blocher says: 'The presence of one or several word-plays casts doubt on any literal intention on the author's part; they reveal an author who is no way naïve, but who uses naïve language for calculated effects. Paul's *ek* [where he states in 1 Corinthians 11:8 that woman was made from (*ek*) man] does not require a literal interpretation of Genesis on this point. There are different kinds of causality, and that which the apostle has in mind may be exemplary or final. It could perfectly well be said that woman is "from" (*ek*) the man if he played the part of a prototype and if God created the woman because of the need the man had of her. Such a conclusion emerges by itself from Genesis 2, even if the text does not reveal the detailed method of divine procedure. The author plays on the double meaning for rib, which also means "side" and therefore "*alter ego*". Arabs use the expression "He is my rib" to mean "He is my close friend". We use a similar turn of phrase when referring to one's "better half". If Paul does not require a literal reading, and if the word for rib/side is rich in symbolism, we have the right to consider the hypothesis of figurative language' (1984: 98–99).

Ingenious commentators have had a field day with symbolic inferences. One rabbinic commentator puts these words into God's mouth: 'Where shall I make her from? Not from the head, lest she stand too proudly; nor from the eyes, lest she be excessively curious; nor from the ears, for she would risk being indiscreet; nor from the nape of the neck, which would only encourage pride . . .' and so on. The rib is selected in order to make the woman modest – but the dreaded faults will appear all the same. The old commentator Matthew Henry, following Thomas Aquinas, is closer to the spirit of Genesis: God did not make the woman from man 'out of his head to rule over him, but out of his side to be equal with him, under his arm to be protected, and near his heart to be beloved'. More subtly, Augustine understood that the man is the strength of the woman (from him comes the *bone*), while the woman softens the man (in the place of the rib, God closes up the *flesh*). There is no reason to attribute all these meanings to the inspired author's intention; but he no doubt had several of them in mind, which increases the probability that the writing is figurative.

7.   The earliest accepted remains of fossil *Homo* (assigned to the species *H. erectus* and found widely distributed in the upper Pleistocene about a million years ago) had a skeleton not unlike our own, but with a smaller brain; well-developed ridges of bone over the eyes; a ridge on the back of the skull to attach powerful neck muscles; a sloping forehead; flat face with no chin; and large upper incisors. Nevertheless *H. erectus* was an erect walker, a meat-eater and toolmaker (belonging to the Acheulian or great hand-axe culture); physically and (in a loose sense) culturally, he was also quite closely related to the australopithecine 'near-men' of Africa. The fossil evidence is supported by an enormous amount of anatomical, physiological and genetical evidence. For example, the chromosomes of humans and the other great apes are almost identical, the main difference being the fusion of two ape chromosomes to form a single element in man; the 'genetic codes' of the DNA of chimpanzee and man differ by less than 2%.

8.   Kidner adds a footnote: 'Gn. 3:20, naming Eve as "mother of all the living". However, the concern of the verse is principally to reiterate, in the context of death, the promise of salvation through "her seed" (Gn. 3:15).'

9.   The link between genes and behaviour received considerable publicity with the publication in 1975 of *Sociobiology* by E. O. Wilson, a Harvard entomologist. Wilson based his arguments on a point made over forty years earlier by J. B. S. Haldane (1932), that if the unselfishness (even to the point of self-sacrifice) of an individual had an inherited basis, and if he (or she) supported his near relatives so that they raised more children than they would otherwise have done, then the altruism genes would be selected and spread in the same way (for example) inherited resistance to a disease. This concept was formalized as 'inclusive fitness'

by W. D. Hamilton in 1964, and by John Maynard Smith in the same year as 'kin selection', the name by which it has become known. The 1950s and 60s saw much interest in biology and behaviour, expressed in the writings of Konrad Lorenz, Niko Tinbergen, Vero Wynne-Edwards, Robert Ardrey and Desmond Morris, later expressed in such television series as David Attenborough's *Life on Earth* (Barnett, 1988). In his book, Wilson reviewed the evolution of social behaviour in many animal groups, and in a concluding chapter extrapolated his conclusions to humankind, claiming that sociobiology, defined as 'an interdisciplinary science which lies between the fields of biology (particularly ecology and physiology) and psychology and sociology', is the key to understanding human nature.

Wilson has been attacked by both sociologists and socialists, who see his ideas as disruptive to their dreams of improving society by manipulating the environment. His suggestions have, however, been attractive to those seeking a naturalistic explanation of human altruism. For example, Peter Singer has written: 'Sociobiology . . . enables us to see ethics as a mode of human reasoning which develops in a group context . . . so ethics loses its air of mystery. Its principles are not laws written up on Heaven. Nor are they absolute truths about the universe, known by intuition. The principles of ethics come from our own nature as social reasoning beings' (1981: 149).

10. In his Word commentary, Gordon Wenham is equivocal: 'I prefer to view Gen. 2–3 as both paradigmatic and protohistorical' (1987: 91).

7. *Evolution*

1. The idea that the Universe as a whole had evolved was not accepted until surprisingly recently, even though Augustine of Hippo had suggested in the fifth century that the *fiat* commands ('Let the earth bring forth . . .' *etc.*) did not refer to once-for-all events, but gave the Earth the power to go on producing new things. In other words, creation was not limited to an initial act, but also to 'the days in which He daily fashions whatever evolves in the course of time'. Modern ideas, however, seem to stem from Immanuel Kant (1724–1804), who, in *A General History of Nature and a Theory of the Heavens* (1755), systematically developed the now familiar idea that the world had its beginning with a chaotic universal nebula that started to rotate and eventually formed the galaxies, suns and planets. Kant emphasized the gradualness of the entire process: 'The future succession of time, by which eternity is unexhausted, will entirely animate the whole range of Space to which God is present, and will gradually put it into that regular order which is comfortable to the excellence of His plan . . . The creation is never finished or complete. It did indeed once have a beginning, but it will never cease' (1755, cited by Mayr 1982: 314).

This was a dynamic, continuously evolving world, remotely governed by secondary causes: 'Kant deliberately set on one side Newton's careful distinction between the creation of the present Order of Nature and its maintenance: the only creation we need demand was the progressive victory of order over chaos through an infinity of time' (Toulmin & Goodfield 1965: 133).

Meanwhile, evidence began to accumulate that the surface of the Earth had not always been the same. One of the first pointers was the discovery of extinct volcanoes in the Pay de Dôme district of central France, which led to the recognition that basalt, a widely distributed rock, is nothing but ancient lava. At about the same time, it was realized that many – probably most – geological strata are sedimentary, deposited from water, and sometimes these deposits may be tens of thousands of metres thick. This was extremely disturbing: it must have taken an immense amount of time for such thick deposits to accumulate, so the Earth must be much older than previously thought. Even worse was the discovery that neither volcanic nor sedimentary rocks remain unchanged after being laid down. They are subsequently eroded as water cuts valleys through them, and in many cases sedimentary layers are folded and occasionally turned over completely at some time after deposition (Gillispie 1951).

All of this stimulated interest in the age of the Earth. Newton – who was a convinced, albeit heterodox, Christian – calculated that the Earth must have cooled for at least 50,000 years until it was cold enough for life, but he felt something must be wrong with his sums because this was so much greater than the church's teaching of 6,000 years since creation.

In *Les Époches de la Nature* (1779), George-Louis Buffon (Director of the Royal Botanic Gardens in Paris 1739–88) reported on some experiments he had done heating a group of spheres of various sizes, from which he concluded that 74,832 years were required for the cooling of the Earth from white heat to its present condition. (He privately estimated that the Earth was at least half a million years old, but he did not publish this figure, because an earlier book of his had been censored.) In an attempt to harmonize his results with Scripture, Buffon argued that there had been seven epochs of Earth history, more or less matching the days of creation in Genesis. In other words, he suggested his interpretation was not all that different from the traditional one, so long as the Genesis days were taken as 'epochs' rather than twenty-four-hour periods: (1) formation of the Earth and planets; (2) origination of the great mountain ranges; (3) water covering all dry land; (4) beginning of volcanic activity; (5) elephants and other tropical animals inhabiting the north temperate area familiar to Buffon; (6) separation of continents (Buffon appreciated the similarity of North American animals to those of Europe and Asia, and reasoned that the two land masses must have been connected with each other at some time in the past); and (7) appearance of man.

Another problem for the eighteenth-century understanding of the Genesis story was the ever-increasing knowledge of fossils. Fossils were, of course, well known, but for a long time they were believed to be nothing more than an accident of nature (*lusus naturae*) rather than the remains of once-living creatures. As time passed, their organic origin seems to have been increasingly accepted and it became commonly believed that they represented creatures drowned in Noah's flood. This was not new; in early Christian times Tertullian (160–225) wrote of fossils in mountains as demonstrating a time when the globe was overrun by water, although it is not clear whether or not he was talking about Noah's time. Both Chrysostom and Augustine thought of the Flood as being responsible for fossils, and Martin Luther was even more certain. Notwithstanding, there were two problems about the 'flood theory'.

First, unknown and – as knowledge of living animals and plants increased apace – presumably extinct organisms were found as fossils. The discovery of extinct organisms conflicted not so much with the Bible as with the theologians' 'principle of plenitude', which held that God in the breadth of his mind had surely created any creature that was possible; and conversely God in his benevolence could not possibly permit any of his own creatures to become extinct. Plenitude was usually linked with the ideas of a *scala naturae*, that is, there could be no gaps between forms in the chain. It also commonly involved an assumption of increasing perfection: more 'soul', more consciousness, more ability to reason, or greater advance towards God. Notwithstanding, extinctions were a problem for interpretation rather than a true conflict with Scripture.

Secondly, information collected by an English surveyor, William Smith, and a French zoologist, Baron Cuvier, indicated that particular rock strata have distinctive fossils. Smith was involved in canal-building and attempting to trace coal seams in mines, and he realized that geological strata could be identified by the fossils they contained. Such strata can sometimes be followed for hundreds of miles, even when the rock formation changes. Smith developed these principles between 1791 and 1799, although his 'stratigraphic map' of England and Wales was not published until 1815. During the same period, French naturalists were actively collecting fossils in the limestone quarries around Paris, and Cuvier worked out the exact stratigraphy of these fossils (mainly mammals) in detail. The conclusion, unpalatable as it was at the time, was that there is a time sequence involved in the laying down of fossil-bearing strata, and that the lowest strata are the oldest. Later on it was possible to correlate strata, not only across England or Western Europe, but between different parts of the world, if allowance is made for the same kind of regional differences which exist today in living faunas and floras.

2. For example, if one species evolves into another, why are forms linking the two species not found? Darwin had to discuss this because of the hangover from Lamarckian speculation that existed when he was writing. Lamarck had suggested that evolutionary change comes from the use or disuse of organs and traits, so that transmutation arises from a varying response in an existing group. It would follow from this that there are no firm limits to any species, nor is a 'natural' classification realistically possible, since the evolving unit is the individual. Darwin rejected this idea of species from his personal study of species in nature, and assumed a definition close to the modern one, that a species is an effectively isolated population or group of populations. He recognized, first, that closely related forms are likely to compete for the same resources, leading to the less favoured one(s) becoming extinct. This has been shown repeatedly by experiment (perhaps most exhaustively in the flour-beetle *Tribolium*), with the important qualification that different varieties (or species) can survive together only where a heterogeneous environment allows different varieties to occupy different niches. The key point, however, is that one form will normally become extinct if a fitter form competes with it.

Secondly, in a large area, different species replace each other geographically. In most cases these species seemed to have evolved in isolation and then expanded their ranges to come into contact. Geological changes (separating or connecting tracts of land) play a part, but at the level of modern species the most important factor has been the Pleistocene (Ice Ages), when previously widely distributed forms were isolated in warmer or wetter refuges for relatively long periods, and changed sufficiently to remain distinct when the climate improved and they were able to reoccupy their former territory. Well-worked examples of this process are the Palaearctic ring of *Larus* gull species, and the *Heliconius* butterflies (and their mimics) in the Amazon basin. It is interesting that Darwin recognized the dynamic influence of historical events in forming new species, in contrast to the implicit assumption both of his contemporaries and of many recent biologists that environments are stale and improbably homogeneous.

Thirdly, transitional forms will almost certainly be less common than either the ancestral or descendant form, and hence liable to be overlooked or to become extinct. As collecting and the description of variation has progressed, it has been recognized that some species previously collected only from widely separated areas may in fact be connected by intermediate forms. We now recognize *clines* of change in particular traits, and also that a species may be *polytypic*, that is, it may contain several geographically separated forms which interbreed to a limit extent. Linnaeus confused matters by giving the name 'variety' indiscriminately to geographical races, domesticated races, non-genetic variants, and inherited 'sports'. The idea of a polytypic species radically alters the Linnaean concept of a species: the older idea was of a species characterized by a gap separating it from other groups, while a polytypic species is defined by actual or potential genetic continuity between allopatric, or geographically separated, populations. This does not, however, affect Darwin's point that groups in the process of change are likely to be uncommon. The discovery of active formation of new species (for example, *Drosophila* flies in Hawaii and cichlid fishes in Lake Victoria), the rarity of transitional forms, and the study of hybrid zones are throwing a great deal of light on the nature and integrity of species. (For general accounts of the genetics of speciation, see Otte & Endler 1989; Skelton 1993).

3. Eldredge & Gould 1972; for a critique see Larson's chapter (23) in Otte & Endler 1989.

4. A dialogue between Richard Dawkins and Michael Poole was published in *Science and Christian Belief* 6: 41–59 (1994); 7: 51–58 (1995). Poole argues that the anti-Christian statements of Dawkins are levelled against a deist 'great watchmaker', not the God revealed in the Bible.

8. Biblical portraits of human nature

1. The unbiblical nature of dualistic views of the human individual are taken for granted today in theological writing such as Jewett 1971, especially his discussions of 'body' (*sōma*) (1971: 201–304), 'life' or 'soul' (*psychē*) (334–357) and 'mind' (*nous*) (358–390). Bultmann (1952) likewise insists that Paul does not set body 'in contrast with the "inner soul" ' since 'man does not *have* a *sōma*; he *is*

sōma' (1952: 194). He cites Rom. 12:1, where 'present your bodies' equals 'yourselves'. He writes: 'Man is called *sōma* in respect of his being able to control himself and be the object of his own action . . . (1 Cor. 9:27) . . . he can yield *himself* to the service of sin or of God (Rom. 6:12 and following) . . . he can expend *himself* for Christ (Phil. 1:20).' Bultmann concludes that such anthropological terms as *sōma, psychē,* and *pneuma* (spirit) denote 'the various possibilities' of action or being; 'man does not consist of two parts, much less of three . . . man is a living unity' (1952: 209). Whiteley, while agreeing that Paul holds 'a unitary view of man', nevertheless notes that 'on rare occasions', such as the experience of death, 'he employs dualistic language' (1974: 39).

2. We are creatures trapped in time. Our personal histories in this space-time are vitally important. As MacKay observed, concerning discussions of enduring personal identity: 'What it does underline, I think, is the importance of an individual's history, his world-line throughout space and time in the definition of his identity. In the relativistic physics (which we don't need to worry about in technical detail) the realistic picture of the physical universe is held to be not the ordinary three-dimensional snapshot that we might take with a camera, but rather a four-dimensional picture in which time is one of the dimensions; popularly speaking and you and I and every object in the world are represented by an extended worm-like object through time in the four-dimensional space-time. The notion of a world-line, as a more unitary representation of entities in the world, is already familiar in that context, and I am suggesting that in the case of us as conscious agents, it is our world-line, our history, which must be pointed to in order to establish our identity.' He continues later: 'What does theism claim as to the nature of reality, all reality? What it claims is that there is only one Reality, namely God the Author, and what the Author brings into being by His creative word, His "Let there be". For theism, the world of physical reality – the succession of events in and through which we encounter what we call the physical world – is a succession of events each of which owes its being to the creative will of an Author. That is the view that theism sets as the context of which it also claims that death need not be the end of us. God, as the Author of all that is, is the final definer of the identity of all that is. He is not merely the author in the sense of an arbitrator. He is the Giver of being to the identity of all that is.'

MacKay concludes his discussion helpfully as follows: 'If I understand it right, however, the world of the resurrection that is envisaged by Christian theism is not to be reduced to just "more of the same" at a later point in time. I see no basis for insisting that that is the thought model to which the Christian hope is pointing, even though a lot of metaphors are used which would fit in with that. But there is also the insistence on the concept of a "New Creation". Here we are in deep water theologically, but at least if the concept of creation is to be thought of by any analogy with creation as we ourselves understand it – as, for example, the creation of a space-time in a novel – then a new creation is not just the running on and on of events later in the original novel: it is a different novel. A new creation is a space-time in its own right. Even a human author can both meaningfully and authoritatively say that the new novel has some of the same characters in it as the old. The identity of the individuals in the new novel is for the novelist to determine. So if there is any analogy at all with the concept of a new creation by our divine Creator, what is set before us is the possibility that in a new creation the Author brings into being, precisely and identically, some of those whom He came to know in and through His participation in the old creation' (1991: 266–267).

9. Human nature: biology and beginning

1. This long-standing tradition was based on Greek embryology as interpreted by the early Christian fathers, notably Augustine, and confirmed by Bible exposition (Dunstan 1984). Aristotle regarded the earliest embryonic stages as plant-like, needing and drawing nourishment like any other plant; the embryo then became an animal creature, sensitive and responsive like other animals; and only thereafter a creature having fully human characteristics. This understanding saw the outward form as the projection of inward animating principles (Greek *psychē*, Latin *anima*, English soul): animating the plant was a vegetative soul, animating the animal was a sensitive soul, and animating the human form was a rational soul. It seemed to accord with Ex. 21:22–23, which is the only reference in the Bible to abortion. This passage clearly affirms a worth for foetal life, but its application is not straightforward. The Septuagint translation says that life is to be given for life if the embryo is 'formed'. Both Jerome and Augustine explained that the act was not to be taken to be homicide if the foetus was 'unformed', 'for there cannot yet be said to be a live soul in a body which lacks sensation when it is not formed in flesh and so not yet endowed with sense' (Augustine on the Latin text of Ex. 21:22). While both Jerome and Augustine remained agnostic about (in traditional language) the point at which the soul entered the body, they were not prepared to affirm with confidence that this had taken place while the foetus was 'unformed'.

The modern use of these verses is not helped by versions which translate the Hebrew (which speaks, literally, of 'departing fruit') as miscarriage (RSV, NEB) or the more emotive 'gives birth prematurely' (NIV). Notwithstanding modern translations, for almost 1,500 years the dominant Christian tradition gave full protection only to a 'formed' embryo. Until the embryo began to have human form, it lacked human animation; in the western moral and legal traditions, *formatus* and *animatus* became criteria for the recognition of a being to be protected. In practice, the time of transition from unformed to formed was taken as 'quickening', when a mother begins to feel her foetus moving. For example, Thomas Aquinas wrote of Ex. 21:22–23, 'If death should result either for the woman or for the animated foetus (*puerperii animati*), he who strikes cannot escape the crime of homicide . . . The conception of a male is not completed until about the fortieth day, as Aristotle says; that of a female until about the ninetieth day.' Scriptural support for the idea that the soul enters a male foetus at forty days and a female at ninety was claimed in papal edicts in the thirteenth and sixteenth centuries, on the basis of the periods of ritual cleansing after childbirth laid down in Lv. 12:1–5.

Embryologists do not now recognize a major distinction between a formed and an unformed foetus, and many modern expositors assume therefore that the traditional understanding is wrong. The point is worth making, however, that the longest-lasting Christian tradition has been that early foetuses do not warrant the same degree of protection as later ones, and that this distinction was based firmly on a respect for and application of Scripture. It is also worth noting that the official (and current) Roman Catholic Declaration on Abortion (1974) is explicitly agnostic about the status of the earliest embryos: 'This declaration deliberately leaves aside at what moment in time the spiritual soul is infused. On this matter tradition is not unanimous and writers differ. Some assert it happens at the first instance of life, while others consider that it does not happen before the seed has taken up its position.' Moreover, Coghlan (1990) has argued that the Vatican's *Instruction on Respect for Human Life in its Origin and on the Dignity of Procreation* (1987), which declares that the embryo is a 'person' from the moment of conception (see below, note 2) is an invalid departure from Roman Catholic moral reasoning, since it cannot be supported independently of religious belief and thus cannnot be a universally acceptable principle.

2. In a review of the relevant biblical data, Rogerson concluded that 'even if we assume that the "image" is asserting something ontological about mankind, what we do not know is whether the "image" (whatever it is) is present from the moment of conception or whether, in Old Testament terms, it is there only after the "unformed substance" has reached its definite human form . . . Nothing in the Bible clearly shows that the image of God is "present" from the moment of conception' (1985: 85).

3. For a general review of such issues see Jones 1984; 1987; Dyson 1995.

4. For example, O'Donovan (1984: 50), while accepting that the expression 'personhood' is not found as such in the Bible, has argued its importance as defining something which is disclosed rather than conferred, and which we recognize only because of a prior moral commitment. Certainly it is implicit in places where the scriptural text refers to the individual man or woman. But it is dangerous to build too much on it in the context of prenatal life. Indeed, Mahoney has described it as 'perhaps the most ambiguous term in the whole discussion . . . the idea of a person (as existing from the fertilization of the ovum) has moved so far from ordinary usage and become so attenuated and blurred, so distanced from ordinary understanding as to be now altogether meaningless, and indeed misleading . . . The debate is not clarified by introducing the idea of potential, since its use on both sides simply reflects and extends (opposing) fundamental positions. Those who would argue that to describe the conceptus as only potentially a human person is to ignore the fact that even the child at birth is still only potentially a human person are using the term 'potential' to mean the capacity to become more of a person, or more fully a person, in terms of characteristically personal activities. While on the other hand, those who claim that this description of a potential human person applies exclusively to the embryo or foetus at an early stage are using the term 'potential' to mean that it is not yet in any real sense a person at all' (1984: 54–55).

5. It is not accurate to speak of genes as being intrinsically 'bad' or 'deleterious', but only to recognize that a characteristic may be harmful *in a particular environment*. Short-sightedness is a problem to a hunter, but can be repaired by spectacles; aggressiveness may be a problem in an urban, sedentary environment, but may have been necessary to survival in a less regulated society. It is easy to multiply such examples: the point to emphasize is that our genetical make-up is not *per se* all-determining. Rather, the person forged from those genes is inseparable from his history. The danger before all of us is failing to attain maturity because we have not exercised our ability to choose our environment.

10. Brain, mind and behaviour

1. Dostoevsky was himself an epileptic, and offers his own experience in the thoughts of Mishkin in *The Idiot*: 'Thinking about this moment afterwards, when he was again in health, he often told himself that all these gleams and flashes of superior self-awareness and, hence, of a "higher state of being", were nothing other than sickness . . . and so, were not the high state of being at all but on the contrary had to be reckoned as the lowest.' And yet he came finally to an extremely paradoxical conclusion. In an essay, 'What if it is sickness?', he asked himself, 'What does it matter if it is abnormal in intensity, if the result, if the moment of awareness, remembered and analysed afterwards in health, turns out to be the height of harmony and beauty, and gives an unheard-of and until then undreamed-of feeling of wholeness, of proportion, of reconciliation, and an ecstatic and prayer-like union in the highest synthesis of life . . . If in that second – that is, in the last lucid moment before the fit – he had time to say to himself clearly and consciously: "Yes, one might give one's whole life for this moment!" Then that moment by itself would certainly be worth the whole life.'

2. In contrast, Persinger has argued that many sudden conversions are the results of abnormal circumstances leading to ecstatic experiences associated with a malfunctioning brain. He observed that the incidence of reported mystical religious experiences in association with seizures in temporal-lobe epilepsy (variously labelled 'complex partial epilepsy' or 'psychomotor epilepsy') is well documented in the clinical literature. The focus of the seizure activity usually begins in the brain stem, and it is for this reason that the auras experienced are associated with dramatic changes in mood, felt variously as euphoria or anxiety or fear. In addition, there are frequently feelings of strangeness, hallucinations and delusions. The combined effect of these events are endowed by the patients with a deep personal significance. For this reason it is not surprising that the overall events take on a strong religious flavour.

Persinger and his colleagues have gone on to explore more or less religious people in the normal (seizure-free) population. They suggest that normal people who report mystical or paranormal religious experiences are in fact suffering 'micro-seizures', *i.e.* slightly abnormal electrical discharges of the limbic, emotional brain, sufficient to produce a mystical experience.

Persinger constructed a personality inventory designed to measure the prevalence of temporal-lobe seizure-like experiences in normal individuals; he included items designed to register any experiences resembling those documented in seizures associated with temporal-lobe epilepsy, the so-called Complex Partial Epileptic Signs (CPES) which seemed to be another name for temporal-lobe epilepsy. He found that college students who reported 'paranormal' (mystical, with religious overtones) experiences and 'a sense of presence' scored highly on his CPES questionnaire and also had a relatively high incidence of abnormal temporal-lobe brain-wave activity. Persinger concluded that there is a continuum of functional hyperconnectivity between limbic and cortical areas in 'non-seizure' patients and that this predisposes some such individuals to mystical and religious experiences.

Persinger's approach is worth pursuing, albeit with caution. In particular, the experiences of seizure patients are almost always bizarre and out of the ordinary, and quite unlike those habitually reported as typical of Christian conversions. Typically, the latter are described as conscious, deliberate decisions often made in non-emotional circumstances and with none of the accompanying bright lights and mystical overtones of epileptic patients. We simply note here that the God of the Christian is just as capable of communicating through normal brain activity as through abnormal brain activity, and indeed would seem to do so habitually. Warren Brown and Carla Caetano comment on Persinger's and related studies: 'Religious experience associated with brain seizures (or drugs), although undeniably a contributor to the conversions of some individuals, have a number of weaknesses as a general neuropsychological model of religious conversion. Most importantly, the accounts of mystical/religious experiences in the clinical epilepsy literature are not characteristic of *typical* Christian (or other) religious conversions. Seldom do individuals describe their religious conversion in terms confusable with seizures, *i.e.* as based on paranormal experiences having a sudden onset and perceived as discontinuous with one's ongoing stream of consciousness.

'The proposition that religious experiences in the normal population are related to "micro-seizures" seems to rest on a tautology, *i.e.* unobserved "micro-seizures" are hypothesized on the grounds that individuals with paranormal religious experiences score high on the CPES, which itself includes questions regarding significant paranormal experiences. There is little evidence to directly support the notion that the CPES is diagnostic of subclinical temporal-lobe seizure events among apparently normal individuals. The Personal Philosophy Inventory and CPES are more likely to be simply different ways of asking individuals about unusual religious experiences.

'From a theological point of view nothing is accomplished by establishing that a particular conversion may have been related to abnormal brain activity; that is, the epistemological problem of the *truth* of the content of the experience is not solved. Imagine, for example, that you have the strong impression that someone is standing behind you but also know that you are having a seizure. It is still necessary to confirm or disconfirm the possibility that someone is there by turning and looking. The fact of the seizure does not bear on the fact that someone may well be standing behind you. In the case of St Paul, if the Damascus road experience happened to involve a seizure it would have little relevance to

the theological question of the truth of what he preached, taught and wrote'
(Brown & Caetano 1992).

3. Professor Andrew Sims, a former President of the Royal College of Psychiatrists
    in Britain, has commented, 'Rather belatedly, the Royal College of Psychiatrists
    has recognized the need to consider spiritual issues, with the College Trainees
    Committee leading the way by recognizing the need to emphasize the physical,
    mental and spiritual aspects of healing in the training of doctors in general and
    psychiatrists in particular. Religious and spiritual factors influence the experience
    and presentation of illness . . . Gellner (1987) has argued cogently that the
    psychiatrist cannot escape society thrusting a priestly role upon her. If these
    expectations do exist, we should quite clearly state that we will not fulfil them. We
    need to balance the importance of the spiritual in the life of our patients with
    denying absolutely any sort of priestly role for ourselves as psychiatrists' (Sims 1994).

11. Psychology

1. Typically an undergraduate psychology student will be taught certain generaliza-
    tions about psychotherapy: (1) There are more than 250 types of psychotherapy
    on offer. (2) Half of all psychotherapists describe themselves as eclectic. (3) There
    are a limited number of basic theoretical perspectives on which the aims and
    techniques of particular therapies are built: psychoanalytic, humanistic,
    behavioural, cognitive and interpersonal. (4) It is extremely difficult to evaluate
    the effectiveness of psychotherapy. How shall it be done? Should it depend on
    the patient's feelings about their progress? On how the therapist feels about their
    progress? On how friends and family feel about it? On how another clinician
    acting independently judges the progress? Or how the patient's actual behaviour
    has changed? What do you assess as an outcome? This, in turn, reflects the presumed
    theory or value underlying this assessment. (5) There are some accepted general
    conclusions about the effectiveness of therapies: (a) people who remain untreated
    often improve (spontaneous remission); (b) those who receive psychotherapy are
    somewhat more likely to improve than those who do not, regardless of what kind of
    therapy they receive and for how long; (c) mature, articulate people with specific
    behaviour problems often receive the greatest benefits from therapy; and (d) placebo
    treatments or the sympathy and friendly counsel of para-professionals also tend to
    produce more improvements than occur in untreated people.

2. Van Leeuwen quotes with approval McKeown's exposure of the way in which
    some Christians have 'over-translated' the Bible into behaviourist language: 'In
    such an analysis, eschatological doctrine is equated with positive reinforcement,
    the conscience functions as a negative reinforcer, and the beatitudes are analogous
    to a system of token rewards. Indeed, it is asserted that Hebrews 11:6 ("Without
    faith, it is impossible to please God, because anyone who comes to him must
    believe that he exists and that he rewards those who earnestly seek him")
    establishes God as a positive reinforcer. At issue in this translation, more than its
    being trite and superficial (comparable to declaring God as my "co-pilot" or
    "cosmic power") is that it *legitimatizes the fiction of behaviourist metapsychology*, and
    in so doing debases religious language and symbolism and sterilizes, if not

destroys, traditional religious images, meanings and significance. The translation is dangerous because it substitutes a new set of symbols for those essential to historic Christian faith' (McKeown 1981; our italics).

3. Gordon Allport had a major influence on the development of theories of personality. He wrote that 'Different as are science and art in their axioms and methods, they have learned to co-operate in a thousand ways – in the production of fine dwellings, music, clothing, design. Why should not science and religion, likewise differing in axioms and method, yet co-operate in the production of an improved human character without which all other human gains are tragic loss? From many sides today comes the demand that religion and psychology busy themselves in finding a common ground for uniting their efforts for human welfare' (1950: viii).

Frederic Bartlett, one of the pioneers and architects of the cognitive revolution in psychology, wrote: 'It is inevitable that the forms which are taken by feeling, thinking, and action within any religion should be moulded and directed by the character of its own associated culture. The psychologist must accept these forms and attempt to show how they have grown up and what are their principal effects. Should he appear to succeed in doing these things, he is tempted to suppose that this confers upon him some special right to pronounce upon the further and deeper issues of ultimate truth and value. These issues, as many people have claimed, seem to be inevitably bound up with the assertion that in some way the truth and the worth of religion come from a contact of the natural order with some other order or world, not itself directly accessible to the common human senses. So far as any final decision upon the validity or value of such a claim goes, the psychologist is in exactly the same position as that of any other human being who cares to consider the matter seriously. Being a psychologist gives him neither superior or inferior authority' (1950: 3–4).

4. In addition to her concerns about the anti-psychologists, Vande Kemp also has unease about what she calls the 'simplistic epistemology' of some Christian psychologists. Vande Kemp's own description of the scientific enterprise and the way scientists work is very similar to that outlined in chapter 3. Thus she writes: 'Scientists rely heavily on the testimony of past witnesses, as any introductory textbook will attest. Most scientists also rely on intuition, as a source of hypotheses or a creative explanation of unexpected findings, and in their development of new instruments. All contemporary scientists integrate reason and observation in the hypothetico-deductive method, in which hypotheses developed on the basis of observation are tested empirically (or experimentally) and used to build models and theories' (1987: 20).

Some Christians are deeply concerned about the seeming exclusive adherence of some psychologists to the experimental method and to a positivistic view of science. They are anxious that something should be done to 'humanize' psychology. Mary Stewart van Leeuwen criticizes such an approach, since 'Human action, they say [*i.e.* the humanizers], cannot be understood merely by observation and description from an outsider's point of view ... methods are needed which will enable the scientists to understand, in active co-operation with the subjects, how the subjects see their particular situation' (1985: 73–74). Foster and Ledbetter agree: 'Van Leeuwen (1985) does not advocate a complete abandonment of the scientific method, but rather believes that we must be willing to modify our procedures to allow the person being studied to be more human', which in practice means relying much more heavily on subjective reports. They go on: 'We can accept this type of research within a broad definition of empiricism, but generalizations from these results would have to be severely limited due to the subjectivism introduced by the procedures' (1987)

Vande Kemp has her own preferred solution, which she calls 'philosophizing psychology . . . a psychology which has no connections with philosophy and theory may not be capable of humanizing, despite our best intentions – and philosophy may be reluctant to welcome us back unless we broaden our thinking (and method) considerably' (1987).

Support for this comes from Michael Shepherd: '. . . both psychiatry and psychology have attempted to sever their conjunction with formal philosophy, basing their claims to independence on the spirit of empiricism and scientific enquiry and condemning, in the words of a major textbook of psychiatry, "attempts to solve the problems of psychopathology by philosophical short-cuts, instead of the relatively slow method of investigation with the disciplines of natural science"' (1995: 287).

Shepherd points out 'that no "philosophical short-cuts" are to be found in the work of the two outstanding medically qualified men who can be credited with both an intimate knowledge of the psychological sciences and a worldwide reputation as philosophers. They reached the same conclusion in slightly different ways. To the psychologist/philosopher William James psychology was "the ante-room to metaphysics" (James 1908) and metaphysics was no more than "an unusually stubborn effort to think clearly". To the psychiatrist/philosopher Karl Jaspers,". . . a thorough study of philosophy is not of any positive value to psycho-pathologists, apart from the importance of methodology . . . but philosophical studies can protect us from putting the wrong question, indulging in irrelevant discussions and deploying our prejudices"' (Shepherd 1995: 287).

The major contribution that philosophy in this sense can make to the science and practice of psychology (and psychiatry) is as a constant reminder of the often hidden presuppositions influencing the questions being asked, and an enabler for selecting the methods appropriate to answering those questions and interpreting the results of any empirical/experimental projects carried out. Thus Aubrey Lewis, one of the leading British psychiatrists of the twentieth century, wrote, 'Nobody in psychiatry can do without a philosophical background, but very often it is an implicit and not an explicit one. This matter has received much less attention than it deserves. Philosophical influences, social influences, religious influences, ideological influences, all play their part in moulding the mental outlook of psychiatrists' (1991). We would add that precisely the same statement could be made of psychologists.

Any closer links between philosophy and psychology depend on psychologists becoming more aware of the history of their discipline and better acquainted with relevant aspects of philosophical writings. Conversely, such links will, as

Francis Crick has pointed out, impose demands on philosophers; a casual acquaintance with what is happening in psychology and neuroscience will not suffice. A sustained detailed knowledge will be required if the views of the philosophers (and theologians) are to be taken seriously by psychologists, psychiatrists and neuroscientists.

This should be obvious to those involved because there is a substantial overlap between religion and psychiatry. The problem is how to understand this. Stanton Jones believes that 'the explicit incorporation of values and world views into the scientific process will not necessarily result in a loss of objectivity or methodological rigor' (1994). There is a difficulty in interpreting what he means by 'incorporation', since it would be regarded as an assertion with which many Christians who are natural scientists would wish to take issue. He is on safer ground when he continues: 'What is new about this proposal . . . is that psychological scientists and practitioners be more explicit about the interaction of religious belief and psychology . . . If scientists, especially psychologists, are operating out of world-view assumptions which include the religious, and the influence of such factors is actually inevitable, then the advancement of the scientific enterprise would seem to be facilitated by making · those beliefs explicitly available for public inspection and discussion.' But why, we may ask, is the influence of such factors actually inevitable?

Jones's main point, however, is that it is wrong to infiltrate personal metaphysical beliefs into psychology, and we fully agree. It was wrong for Skinner to interpret his approach to psychology as if it were an intrinsic part of the discipline rather than metaphysical beliefs which he chose to bring to it; it was wrong for Mallow to present his theories as if his personal views were an intrinsic part of the theory and flowed from the data. But if it was wrong for these psychologists to 'incorporate' their metaphysical beliefs and worldviews as they did, the question arises: why is it right for Christians to 'incorporate' *their* beliefs?

12. *Our common future*

1. 'Without agape, human love for nature will always be dominated by unrestrained eros, and will always be distorted by extreme self-interest and material evaluation (which results in acquisitiveness). Not only can agape transform eros, but it can also provide an eschatological vision of nature quite independent of our day-to-day needs' (Bratton 1992: 15).

2. Lecture given at St James's Church, Piccadilly, London; see also Fox (1983). Criticisms of Fox's theology appear in Brearley 1992; Bishop 1992.

3. Ernest Lucas writes: 'One of the marks of culture since the Enlightenment has been confidence in the power of analytical reasoning to get to "the truth" about life, the universe and everything. The form of analytical reasoning which came to be adopted as the universal paradigm – to be idolized, one might say – was that used in the physical sciences, or, more correctly, an idealized version of that form. Because the physical sciences are concerned only with matter and energy (which we know are inter-convertible) and with interactions between different forms of matter and energy, they are inherently materialistic, and this limitation

tended to be taken over into other disciplines. Perhaps the clearest example of where this leads was seen in logical positivism . . . [This] has now faded away as a major philosophy, partly because its basic axiomatic statement that "the only meaningful synthetic statements are those that are empirically verifiable" is itself not empirically verifiable. It does, however, illustrate the ethos of Western culture in the twentieth century, in its implicit denigration of matters moral and spiritual. To some extent, the criticisms which New Age adherents make of Western culture echo what some Christians have been saying for a long time. However, Christians need to be discriminating here. In their reaction against Western rationalism, New Agers . . . tend to subordinate rationality to intuition or to mystical ways of knowing . . . A biblical Christianity, while rejecting the idolizing of human reason, leads to a more positive attitude towards the place of rationality in seeking and discovering truth than is found in New Age circles' (1996: 150).

4. The irony is that disturbance of the Gaian system is ineffective, since it would return to its original condition when the disturbance ended. This is rather embarrassing for devout Gaians. See Deane-Drummond 1992; Ruether 1992.

5. This must not be allowed to denigrate the intellectual importance of the Strategy, which established the interdependence of environmental care and development action, and set guidelines for conservation action. It was a seminal document which underpinned subsequent international agreements (Berry 1996a).

6. Unpublished lecture given at Godolphin & Latymer School, London, 1990.

7. For a biblical analysis of the appeal of wilderness, see Bratton 1993.

13. The implications of science

1. It is relevant here to refer again to Frederic Bartlett's Riddell Lectures which he called *Religion as Experience, Belief, Action* (1950). He wrote: 'A religious experience is based upon evidence or information which is collected through the normal exercise of human functions, and is always related to isssues of a personal or social character. These are issues which readily arouse feelings and emotions, hopes and fears, longings and ambitions that are deeply cherished. The evidence appears unfinished. It is completed, not by argument, or by the accepted methods of scientific experiment, but by an intuitive leap. Thus this completion appears to come from outside, and more than that, the religious completion is treated as coming from a world, the constituent items of which do not have the properties of the items of the evidence. Neither system can be translated completely into terms of the other; acceptance alone can bridge the gap. In the religious experience there is much that is imperfectly communicable, or not communicable at all. Its results, as they take place within the natural world, can be described just as other things in the same world. But once the religious intuitive leap is taken and its completion accepted, their explanation claims to lie irrevocably outside that field which can be explored within the realm of the knowledge of natural events.' What Barlett talks about as 'an intuitive leap', in Christian circles we would call commitment or personal faith.

Further reading

Perhaps the best way to get a perspective on the science–faith interface is through the history of the interactions between the two, well explored by Colin Russell (1985) and John Hedley Brooke (1991). It will be clear that our understanding of the situation has developed relatively recently, starting from the problem that the Earth – indeed the whole Universe – could be thought of as a machine, and hence raising the questions: 'Where does God fit in?' and 'How does he relate to creation?' For Paley, there was no difficulty: God was the 'divine watchmaker', and this interpretation was comfortably continued in the first half of the nineteenth century with the Bridgewater Treatises, endowed by a bequest from the Eighth Earl of Bridgewater (1756–1829) to expound 'the Power, Wisdom and Goodness of God, as manifested in the Creation'.

The Darwinian revolution, however, raised further questions about God's actions in his creation, but convincing explanations were slow in appearing. When we were students there were few helpful books on the subject. The most read ones were Arthur Rendle Short's *Modern Discovery and the Bible* (1942) and the various works of Robert Clark (1949; 1960 and others), none of them wholly satisfying to someone with questions, but indicators that science–faith issues were actually on the Christian agenda, and that committed Christians were exploring them. Then Bernard Ramm produced *The Christian View of Science and Scripture* (1954), which was more positive.

Since that time there has been a flood of books, good, bad and indifferent. In the year (1966) of the Oxford conference which led to the first edition of this book, Ian Barbour's seminal survey *Issues in Science and Religion* appeared, supplemented since by his Gifford Lectures, *Religion in an Age of Science* (1990; 1998) and *Ethics in an Age of Technology* (1992). We believe that the most significant books that have appeared over the past few decades are those by Donald MacKay, expounding the concept of God 'upholding' all things all the time. We would also commend those of John Polkinghorne, formerly Professor of Mathematical Physics in the University of Cambridge. A number of collected readings have been produced, and several bibliographies. The American Scientific Affiliation (Ipswich, MA) produces *Contemporary Issues in Science and Faith: An Annotated Bibliography* (3rd edition, 1992) and this gives a wider range of reading for anyone wanting to dig deeper and wider than this book.

Bibliography

Allen, D. (1989). *Christian Belief in a Postmodern World*. Louisville, KY: Westminster.

Allport, G. W. (1950). *The Individual and his Religion*. New York: Macmillan.

Aquinis, H., & Aquinis, M. (1995). Integrating psychological science and religion (response to S. L. Jones). *American Psychologist* 50:541–542.

Argyle, M. (1951). *Religious Behaviour*. London: Routledge.

Argyle, M., & Beit-Hallahmi, B. (1975). *The Social Psychology of Religion*. London: Routledge & Kegan Paul.

Ashby, E. (1978). *Reconciling Man with the Environment*. London: Oxford University Press.

Assisi Declarations (1986). *Messages on Man and Nature*. Gland, Switzerland: Worldwide Fund for Nature.

Atkins, P. (1994). *Creation Revisited*. Harmondsworth: Penguin.

Atkinson, D. (1990). *The Message of Genesis*. Leicester: IVP.

—— (1993). Towards a theology of health. In Health: The Strength to be Human. 15–38. Fergusson, 1993.

Attfield, R. (1983). *The Ethics of Environmental Concern*. Oxford: Blackwell (revised edition 1991: Athens, GA: University of Georgia Press).

Austin R. C. (1987). *Baptized into Wilderness: A Christian Perspective on John Muir*. Atlanta: John Knox.

Ayala, F. J. (1974). Introduction. In *Studies in the Philosophy of Biology:*

Reductionism and Related Problems: vii–xvi. Ayala, F. J., & Dobzhansky, Th. (eds). London: Macmillan.

Ayer, A. J. (1946). *Language, Truth and Logic*. London: Gollancz.

Bancroft, J. (1994). Homosexual orientation. *British Journal of Psychiatry* 164:437–460.

Banner, M. (1990). Lecture delivered at the Ramsey Centre in Oxford.

Barbour, I. G. (1966). *Issues in Science and Religion*. New York: Prentice-Hall.

—— (1990, 1992). *Religion in an Age of Science; Ethics in an Age of Technology*. London: SCM. (An encyclopaedic overview. 2 vols.; Gifford Lectures, 1989–91).

—— (1998). *Religion and Science: Historical and Contemporary Issues*. London: SCM. (An expanded version of Barbour 1990.)

Barnes, J. (1979). *The Presocratic Philosophers* (London: Routledge & Kegan Paul). 2 vols.

Barnett, S. A. (1988). *Biology and Freedom*. Cambridge: Cambridge University Press.

Barr, J. (1993). *Biblical Faith and Natural Theology*. Oxford: Clarendon.

Barrow, J. D. (1991). *Theories of Everything*. Oxford: Clarendon.

Barrow, J. D., & Tipler, F. J. (1986). *The Anthropic Cosmological Principle*. Oxford: Oxford University Press.

Bartlett, F. C. (1950). *Religion as Experience, Belief, Action*. Oxford: Oxford University Press.

Bear, D., & Feddio, P. (1977). Quantitative analysis of interictal behaviour in temporal lobe epilepsy. *Archives of Neurology* 34:454–467.

Bergin, A. E. (1980). Psychotherapy and religious values. *Journal of Consulting and Counselling Psychology* 48:95–105.

Berry, A. C. (1987). *The Rites of Life*. London: Hodder & Stoughton.

—— (1993). *Beginnings: Christian Views of the Early Embryo*. London: Christian Medical Fellowship.

Berry, R. J. (1975). *Adam and Ape*. London: Falcon.

—— (1982). *Neo-Darwinism*. London: Edward Arnold.

—— (1987). The theology of DNA. *Anvil* 4:39–49.

—— (1988). *God and Evolution*. London: Hodder & Stoughton.

—— (ed.) (1991). *Real Science, Real Faith*. Eastbourne: Monarch.

—— (1993a). Green religion and green science. *Journal of the Royal Society of Arts* 141:305–318.

—— (1993b). Environmental concern. In *Environmental Dilemmas*: 242–264. Berry, R. J. (ed.). London: Chapman & Hall.

—— (1995). Creation and environment. *Science and Christian Belief* 7:21–43.

—— (1996a). *God and the Biologist*. Leicester: Apollos.

—— (1996b). The virgin birth of Christ. *Science and Christian Belief* 8:101–110.

Berry, R. J., Crawford, T. J., & Hewitt, G. M. (eds) (1992). *Genes in Ecology.* Oxford: Blackwell Scientific.

Birch, L. C. and Cobb, J. B. (1981). *The Liberation of Life.* Cambridge: Cambridge University Press.

Birch, L. C., Eakin, W., & McDaniel, J. B. (eds) (1990). *Liberating Life. Contemporary Approaches to Ecological Theology.* Maryknoll, NY: Orbis.

Bishop, S. (1992). A fox in sheep's clothing. *Third Way* 14. 10:16–18.

Blamires, H. (1963). *The Christian Mind.* London: SPCK.

Blocher, H. (1984). *In the Beginning.* Leicester: IVP.

Boswell, J. (1980). *Christianity, Social Tolerance and Homosexuality* (Chicago: University of Chicago Press).

Bourdeau, Ph., Fasella, P. M., & Teller, A. (eds) (1990). *Environmental Ethics.* Luxembourg: CEC.

Braithwaite, R. B. (1953). *Scientific Explanation.* Cambridge: Cambridge University Press.

Bratton, S. P. (1992). Loving nature: eros or agape? *Environmental Ethics* 14:3–25.

—— (1993) *Christianity, Wilderness and Wildlife.* London and Toronto: Associated University Press.

Brearley, M. (1992). Matthew Fox and the Cosmic Christ. *Anvil* 9:39–54.

Brooke, J. H. (1991). *Science and Religion: Some Historical Perspectives.* Cambridge: Cambridge University Press.

—— (1992). Natural law in the natural sciences: the origin of modern atheism? *Science and Christian Belief* 4:83–103.

Brown, W. S., & Caetano, C. (1992). Conversion, cognition and neuropsychology. In *Handbook of Conversion*: 147–158. Malony, H. N., & Southard, S. (eds). 147–158. Birmingham, AL: Religious Education Press.

Brunner, E. (1939). *Man in Revolt.* Guildford: Lutterworth.

Bube, R. H. (1995). *Putting It All Together.* Lanham: University Press of America.

Bultmann, R. (1952). *Theology of the New Testament.* New York: Scribner.

Byrne, R. (1995). *The Thinking Ape.* Oxford: Oxford University Press.

Calvin, J. *Institutes of the Christian Religion* (1536 onwards), various editions.

Cameron, N. M. de S. (1983). *Evolution and the Authority of the Bible.* Exeter: Paternoster.

Caring for the Earth (1991). *A Strategy for Sustainable Living.* Gland, Switzerland: International Union for the Conservation of Nature, United Nations Environmental Programme, World Wide Fund for Nature.

Carter, B. (1974). In *Confrontation of Cosmological Theories with Observational Data*: 291. Longair, M. S. (ed.). Dordrecht: Reidel.

Christians and the Environment (1991): *A Report by the Board for Social Responsibility.* London: General Synod. Miscellaneous Paper 367.

Clark, R. E. D. (1949). *The Universe: Plan or Accident?* London: Paternoster.

—— (1960). *Christian Belief and Science: A Reconciliation and a Partnership.* London: English Universities Press.

Cobb, J. B. (1969). *God and the World.* John Knox.

Coghlan, M. J. (1990). *The Vatican, the Law, and the Human Embryo.* Basingstoke: Macmillan.

Cole, J. R. (1983). Scopes and beyond: antievolutionism and American culture. In *Scientists Confront Creationism*: 13–32. Godfrey, L. R. (ed.). New York: Norton.

Coulson, C. A. (1955). *Science and Christian Belief.* Oxford: Oxford University Press.

Cranfield, C. E. B. (1974). Some observations on Romans 8:19–21. In *Reconciliation and Hope: New Testament Essays on Atonement and Eschatology presented to L. L. Morris on his 60th Birthday:* 224–230. Banks, R. (ed.). Grand Rapids, MI: Eerdmans.

Crick, F. (1994). *The Astonishing Hypothesis.* New York: Simon & Schuster.

Cupitt, D. (1984). *The Sea of Faith.* London: BBC.

Damasio, A. (1994). *Descartes' Error: Emotion, Reason and the Human Brain.* New York: Grosset/Putnam.

Darley, J. M., & Batson, C. D. (1973). From Jerusalem to Jericho: a study of situational and dispositional variables in helping behaviour. *Journal of Personality and Social Psychology* 27:100–108.

Darwin, C. (1859). *On the Origin of Species by Means of Natural Selection.* London: John Murray.

—— (1871). *The Descent of Man.* London: John Murray.

Davies, G. (1992). *Genius and Grace: Sketches from a Psychiatrist's Notebook.* London: Hodder and Stoughton.

Davies, P. (1983). *God and the New Physics.* Harmondsworth: Pelican.

—— (1992). *The Mind of God.* New York: Simon & Schuster.

Dawes, R. (1994). *House of Cards: Psychology and Psychotherapy Built on Myth.* New York: Free Press.

Dawkins, R. (1976). *The Selfish Gene.* Oxford: Oxford University Press.

—— (1986). *The Blind Watchmaker.* London: Longman.

—— (1995). *River out of Eden.* London: Weidenfeld & Nicolson.

—— (1996). *Climbing Mount Improbable.* London: Viking.

Day, A. (1998). Interpreting the biblical creation accounts. *Science and Christian Belief* (in press)

De Vries, P. (1986). Naturalism in the natural sciences: a Christian perspective. *Christian Scholars' Review* 15:388–396.

De Waal, E. (1991). *A World Made Whole: Rediscovering the Celtic Tradition.* London: HarperCollins.

Deane-Drummond, C. (1992). God and Gaia. *Theology* 95:277–285.

Delacour, J. (1995). The biology and neuropsychology of consciousness. *Neuropsychologia* 33. 9:1061–1192.

Delors J. (1990). Opening address. In Bourdeau, Fasella, & Teller 1990: 19–28.

Desmond, A. (1994). *Huxley: The Devil's Disciple*. London: Michael Joseph.

Dewhurst, K., & Beard, A. W. (1970). Sudden religious conversions in temporal lobe epilepsy. *British Journal of Psychiatry* 117:497–507.

DeWitt, C. B. (ed.) (1991). *The Environment and the Christian*. Grand Rapids, MI: Baker.

Diamond, J. (1991). *The Rise and Fall of the Third Chimpanzee*. London: Vintage.

Doye, J., Golby, I., Line, C., Lloyd, S., Shellard, P., & Tricker, D. (1995). Contemporary perspectives on chance, providence, and free will. *Science and Christian Belief* 7:117–139.

Draper, J. W. (1875). *A History of the Conflict between Religion and Science* London: International Scientific Series.

Duke of Edinburgh & Mann, M. (1989). *Survival or Extinction?* Windsor: St George's House.

Dunn, J. D. (1988). *Romans 1–8*. Word Biblical Commentary. Waco, TX: Word.

Dunstan, G. R. (1983). Social and ethical aspects. In *Developments in Human Reproduction and their Eugenic, Ethical Implications*: 213–226. Carter, C. (ed). London: Academic.

Dunstan, G. R. (1984). The moral status of the human embryo: a tradition recalled. *Journal of Medical Ethics* 10:38–44.

Dyson, A. (1995). *The Ethics of IVF*. London: Mowbray.

Eccles, J. C. (ed.) (1966). *Brain and Conscious Experience*. Berlin: Springer-Verlag.

—— (1970). *Facing Reality*. Berlin: Springer-Verlag.

—— (1989). *Evolution of the Brain*. London: Routledge.

Edelman, G. (1992). *Bright Air, Brilliant Fire: On the Matter of the Mind*. London: Penguin.

Egerton, F. N. (1973). Changing concepts in the balance of nature. *Quarterly Review of Biology* 48:322–350.

Eldredge, N., & Gould, S. J. (1972). Punctuated equilibria: an alternative to phyletic gradualism. In *Models in Paleobiology*: 82–115. Schopf, T. J. M. (ed.). San Francisco: Freeman & Cooper.

Elsdon, R. (1992). *Greenhouse Theology*. Tunbridge Wells: Monarch.

Elton, C. S. (1930). *Animal Ecology and Evolution*. Oxford: Oxford University Press.

Ettlinger, G. (1984). Humans, apes and monkeys: the changing neuropsychological viewpoint. *Neuropsychologia*, 22: 685–696.

288   *Science, life and Christian belief*

Evans, C. S. (1979). *Preserving the Person: A Look at the Human Sciences.* Leicester: IVP.

Fairbairn, A. M. (1902). *The Philosophy of the Christian Religion.* London: Hodder & Stoughton.

Faricy, R. (1982). *Wind and Sea Obey Him.* London: SCM.

Farrer, A. (1964). *Saving Belief.* London: Hodder & Stoughton.

—— (1967). *Faith and Speculation.* London: A. & C. Black.

Fergusson, A. (ed.) (1993). *Health: The Strength to be Human.* Leicester: IVP.

Fisher, R. A. (1954). Retrospects of the criticisms of the theory of natural selection. In *Evolution as a Process*: 84–98. Huxley, J., Hardy, A. C., & Ford, E. B. (eds). London: Allen & Unwin.

Ford, N. M. (1988). *When Did I Begin?* Cambridge: Cambridge University Press.

Foster, J. D., & Ledbetter, M. F. (1987). Wheat and tares: responding to Vande Kemp and other revisionists. *Journal of Psychology and Theology* 15:19–26.

Forster, R., & Marston, P. (1989). *Reason and Faith: Do Modern Science and Christian Faith Really Conflict?* Eastbourne: Monarch.

Fox, M. (1983). *Original Blessing: A Primer in Creation Spirituality.* Santa Fe, New Mexico: Bear & Co.

Furnish, V. P. (1984). *II Corinthians. Translated with introduction, notes, and commentary.* Anchor Bible 32A (Garden City, NY: Doubleday).

Galton, F. (1874). *English Men of Science: Their Nature and Nurture.* London: Macmillan.

Gardner, H. (1992). Scientific psychology: should we bury it or praise it? *New Ideas in Psychology* 19. 2:179–190.

Gillispie, C. C. (1951). *Genesis and Geology.* Cambridge: MA: Harvard University Press.

Godfrey, L. R. (ed.) (1983). *Scientists Confront Creationism.* New York: Norton.

Gottfried, K., & Wilson, K. G. (1997). Science as a cultural construct. *Nature*, (London) 386:545–547.

Gould, S. J. (1981). *The Mis-Measure of Man.* New York: Norton.

—— (1993). Fall in the House of Usher. In *Eight Little Piggies*: 181–193. London: Jonathan Cape.

Granberg-Michaelson, W. (ed.) (1987). *Tending the Garden.* Grand Rapids, MI: Eerdmans.

Green, J. B. (1988). Bodies – that is, human lives: a re-examination of human nature in the Bible. In *Portraits of Human Nature*. Brown, W. S., Murphy, N., & Malony, H. N. (eds). Minneapolis, MN: Fortress.

Greenwald, A. (1984). The totalitarian ego: fabrication and revision of personal history. *American Psychologist* 35:603–613.

Grene, M. (1966). *The Knower and the Known.* London: Faber.

Groothuis, D. P. (1986). *Unmasking the New Age.* Leicester: IVP.

Gross, P. R., & Levitt, N. (1994). *Higher Superstition*. Baltimore: Johns Hopkins University Press.

Grove-White, R. (1992). Human identity and the environmental crisis. In *The Earth Beneath*: 13–34. Ball, J., Goodall, M., Palmer, C., & Reader, J. (eds). London: SPCK.

Grubb, M., Kock, M., Thomson, K., Munson, A., & Sullivan, F. (1993). *The Earth Summit Agreements*. London: Earthscan for the Royal Institute for International Affairs.

Guthrie, W. K. C. (1953). *Greek Philosophy*. Cambridge: Cambridge University Press.

—— (1957). *In the Beginning*. London: Methuen.

Hacking, I. (ed.) (1981). *Scientific Revolutions*. Oxford: Oxford University Press.

Hadamard, J. (1945). *The Psychology of Invention in the Mathematical Field*. Princeton, N.J.: Princeton University Press.

Haldane, J. B. S. (1927). *Possible Worlds*. Chatto & Windus.

—— (1932). *Causes of Evolution*. London: Longmans, Green.

Hall, D. J. (1986). *Imaging God: Dominion as Stewardship*. Grand Rapids, MI: Eerdmans.

Hamilton, W. D. (1964). The genetical evolution of social behaviour. *Journal of Theoretical Biology* 7:1–52.

Hanson, N. R. (1958). *Patterns of Discovery*. Cambridge: Cambridge University Press.

Hardy, A. C. (1975). *The Biology of God*. London: Jonathan Cape.

Hartshorne, C. (1967). *A Natural Theology for Our Time*. La Salle, IL: Open Court.

Hawking, S. (1988). *A Brief History of Time*. London: Bantam.

Hays, R. B. (1986). Relations natural and unnatural: a response to John Boswell's exegesis of Romans 1. *Journal of Religious Ethics* (Spring).

Helm, P. (ed.) (1987). *Objective Knowledge: A Christian Perspective*. Leicester: IVP.

Henry, C. F. H. (ed.) (1978). *Horizons of Science: Christian Scholars Speak Out*. San Francisco: Harper & Row.

Hinde, R. A. (1987). Animal–Human comparisons. In *The Oxford Companion to the Mind*: 25–27. Gregory, R. L. (ed.). Oxford: Oxford University Press.

Hobhouse, L. T. (1913). *Development and Purpose: An Essay towards a Philosophy of Evolution*. London.

Hodge, C. (1874). *What is Darwinism?* New York: Scribner, Armstrong.

Hoekema, A. A. (1986). *Created in God's Image*. Grand Rapids, MI: Eerdmans.

Hooykaas, R. (1957). *Philosophia Libera: Christian Faith and the Freedom of Science*. London: Tyndale.

—— (1971). *Religion and the Rise of Modern Science*, Edinburgh: Scottish Academic Press.

Houghton, J. (1994). *Global Warming: An Investigation of the Evidence, the Implications and the Way Forward*. Oxford: Lion.

—— (1995). *The Search for God: Can Science Help?* Oxford: Lion.

Hoyle, F. (1983). *The Intelligent Universe*. London: Michael Joseph.

Hume, D. (1748/1975). *An Inquiry Concerning Human Understanding*. London. Reissued Oxford: Oxford University Press. 3rd edn.

Hummel, C. E. (1986). *The Galileo Connection: Resolving Conflict between Science and the Bible*. Downers Grove, IL: IVP.

Humphreys, D. R. (1995). *Starlight and Time*. Colorado Springs, CO: Creation Life.

Huxley, A. (1932). *Brave New World*. London: Chatto & Windus.

Huxley, L. (ed.). (1908). *Life and Letters of Thomas Henry Huxley* 3. London: Macmillan.

Huxley, T. H. (1870). *Lay Sermons, Addresses and Reviews*. London: Macmillan.

Huxley, T. L. H. (1887). On the reception of the 'Origin of species' in *Life and Letters of Charles Darwin*: 179–204. Darwin, F. (ed). London: John Murray.

Ison, D. (1983). *Artificial Insemination by Donor*. Grove Booklet on Ethics 52. Bramcote: Grove.

Jacoby, A. (1968). *Señor Kon-Tiki*. London: Allen & Unwin.

James, W. (1902). *The Varieties of Religious Experience*. London: Longmans, Green.

Jantzen, G. (1984). *God's World, God's Body*. London: Darton, Longman & Todd.

Jeeves, M. A. (1969). *The Scientific Enterprise and Christian Faith*. London: Tyndale.

—— (1976). *Psychology and Christianity: The View Both Ways*. Leicester: IVP.

—— (ed.) (1984). *Behavioural Sciences: A Christian Perspective*. Leicester: IVP.

—— (1991). The status of humanity in relation to the animal kingdom. In *Interpreting the Universe as Creation*: 113–122. Brummer, V. (ed.). Kampen: Kok Pharos.

—— (1994). *Mind Fields: Reflections on the Science of Mind and Brain*. Sydney: Lancer; Grand Rapids: Baker; Leicester: Apollos.

—— (1997). *Human Nature at the Millennium*. Grand Rapids: Baker; Leicester: Apollos.

Jegen, M. E. (1987). The church's role in healing the earth. In Granberg-Michaelson, 1987: 93–113.

Jewett, P. (1971). *Paul's Anthropological Terms: A Study of their Use in Conflict Settings*. Leiden: Brill.

Johnson, P. E. (1991). *Darwin on Trial*. Washington, DC: Regnery Gateway.

—— (1995). *Reason in the Balance*. Downers Grove, IL: IVP.

Jones, D. G. (1969). *Teilhard de Chardin: An Analysis and Assessment*. London: Tyndale.

—— (1984). *Brave New People*. Leicester: IVP.

—— (1987). *Manufacturing Humans*. Leicester: IVP.

Jones, J. S., Martin, R. D., & Pilbeam, D. (1992). *The Cambridge Encyclopaedia of Human Evolution*. Cambridge: Cambridge University Press.

Jones, S. L. (1994). A constructive relationship for religion with the science and profession of psychology: perhaps the boldest model yet. *American Psychologist* 49:184–199.

Judge, S. (1991). How not to think about miracles. *Science and Christian Belief* 3:97–102.

—— (1995). Brains and persons. Unpublished paper presented to the Christians in Science annual conference, London.

Kidner, D. (1967). *Genesis*. London: Tyndale.

—— (1975). *Psalms 73–150*. London: IVP.

Kirk, G. S. (1958). Review of *The Physical World of the Greeks*. *Classical Review* 82:111–116.

Kitcher, P. (1982). *Abusing Science: The Case against Creationism*. Boston, MA: MIT Press.

Kitto, H. F. D. (1951). *The Greeks*. Harmondsworth: Pelican.

Kuhn, T. (1970). *The Structure of Scientific Revolutions*. Second Edition. Chicago: Chicago University Press.

—— (1977). *The Essential Tension: Selected Studies in Scientific Tradition and Change*. Chicago: Chicago University Press.

Labuschagne, C. J. (1991). Creation and the Status of Humanity in the Bible. In *Interpreting the Universe as Creation*. V. Brümmer (ed.). Kampen: Pharos.

Lakatos, I. (1970). Falsification and the methodology of scientific research programmes. In *Criticism and the Growth of Knowledge*: 91–95. Lakatos, I., & Musgrave, A. (eds). Cambridge: Cambridge University Press.

Laslett, P. (ed.) (1950). *The Physical Basis of Mind*. London; Macmillan.

Leeuwen, M. S. van (1982) *The Sorcerer's Apprentice: A Christian Looks at the Changing Face of Psychology*. Downers Grove, IL: IVP.

—— (1985). *The Person in Psychology*. Leicester: IVP.

Leopold, A. (1949). *A Sand County Almanac*. New York: Oxford University Press.

Lewis, A. (1991). Dilemmas in psychiatry. *Psychological Medicine* 21:581–585.

Lewis, C. S. (1960). *Miracles*. Rev. edn London: Geoffrey Bles. (1st edn 1947).

Lipowski, Z. J. (1989). Psychiatry: mindless or brainless, both or neither. *Canadian Journal of Psychiatry* 34:249–254.

Livingstone, D. N. (1987). *Darwin's Forgotten Defenders: The Encounter between Evangelical Theology and Evolutionary Thought*. Grand Rapids, MI: Eerdmans.

Loftus, E. (1993). The reality of repressed memories. *American Psychologist* 48:518–537.

Lovelock, J. A. (1979) *Gaia: A New Look at Life on Earth.* Oxford and New York: Oxford University Press.

—— (1988). *The Ages of Gaia.* New York: Norton.

Lloyd, G. E. R. (1970). *Early Greek Science: Thales to Aristotle.* London: Chatto & Windus.

—— (1973). *Greek Science after Aristotle.* London: Chatto & Windus.

—— (1979). *Magic, Reason and Experience: Studies in the Origins and Development of Greek Science.* Cambridge: Cambridge University Press.

—— (1991). *Methods and Problems in Greek Science.* Cambridge: Cambridge University Press.

Lucas, E. C. (1989). *Genesis Today.* London: Scripture Union.

—— (1996). *Science and the New Age Challenge.* Leicester: Apollos.

Lumsden, C. J., & Wilson, E. O. (1983). *Promethean Fire: Reflections on the Origin of Mind.* Cambridge, MA: Harvard University Press.

Lyell, C. (1863). *The Antiquity of Man.* London: John Murray.

McCosh, J. (1896). *The Life of James McCosh: A Record Chiefly Autobiographical.* Edinburgh: T. & T. Clark.

McDonald, H. D. (1981). *The Christian View of Man.* London: Marshall, Morgan & Scott.

McHarg, I. L. (1969). *Design with Nature.* New York: Doubleday.

McIntosh, R. P. (1985). *The Background of Ecology.* Cambridge: Cambridge University Press.

MacKay, D. M. (1953). *The Christian Graduate* 6. 4.

—— (1960). *Science and Christian Faith Today.* London: Falcon.

—— (1974). *The Clockwork Image.* London: IVP.

—— (1978). *Science, Chance and Providence.* Oxford: Oxford University Press.

—— (1979). *Human Science and Human Dignity.* London: Hodder & Stoughton.

—— (1984). The beginnings of personal life. *In the Service of Medicine* 30 (2):9–13.

—— (1988). *The Open Mind and Other Essays.* Tinker, M. (ed.). Leicester: IVP.

—— (1991). *Behind the Eye.* Oxford: Blackwell.

McKeown, B. (1981). Myth and its denial in a secular age: the case of behaviourist psychology. *Journal of Psychology and Theology* 9:19.

McMullin, E. (1993). Evolution and special creation. *Zygon* 28:299–335.

Mahoney, J. (1984). *Bioethics and Belief.* London: Sheed & Ward.

Malinowski, B. (1927). *Sex and Expression in Primitive Society.* London: Kegan Paul.

—— (1936). *The Foundations of Faith and Morals.* London: Oxford University Press.

Mascall, E. (1959). *The Importance of Being Human*. London: Oxford University Press.

Maslow, A. H. (1968). *Towards a Psychology of Being*. New York: Van Nostrand Reinhold.

Maynard Smith, J. (1989). *Evolutionary Genetics*. Oxford: Oxford University Press.

Mayr, E. (1982). *The Growth of Biological Thought*. Cambridge, MA: Harvard University Press.

—— (1991). *One Long Argument: Charles Darwin and the Genesis of Modern Evolutionary Thought*. London: Allen Lane.

Meadows, D. H., Meadows, D. L., & Randers, J. (1972). *The Limits of Growth*. New York: Universe Books.

Medawar, P. (1984). *The Limits of Science*. New York: Harper & Row.

—— (1990a). Is the scientific paper a fraud? In *The Threat and the Glory*: 228–233. Oxford: Oxford University Press.

—— (1990b). Scientific fraud. In *The Threat and the Glory*: 64–70. Oxford: Oxford University Press.

Midgley, M. (1992). *Science as Salvation*. London: Routledge.

Miller, G. (1995). Review of R. Dawes's *House of Cards*. *Psychological Science* 6:129–132.

Milne, A. J. M. (1986). Human rights and the diversity of morals. In *Rights and Obligations in North–South Relations*: 8–33. Wright, M. (ed.). New York: St Martin's Press.

Milne, E. A. (1952). *Modern Cosmology and the Christian Idea of God*. London: Oxford University Press.

Milner, A. D., & Goodale, M. A. (1995). *The Visual Brain in Action*. Oxford: Oxford University Press.

Moltmann, J. (1985). *God in Creation*. London: SCM.

Monod, J. (1970). *Le Hasard et la nécessité*. Paris: Éditions du Seuil. Eng. trans. *Chance and Necessity*. London: Collins, 1972.

Montefiore, H. (ed.) (1975). *Man and Nature*. London: Collins.

Moule, C. F. D. (1964). *Man and Nature in the New Testament*. London: Athlone.

—— (1965–66). St Paul and dualism: the Pauline conception of resurrection. *New Testament Studies* 13:106–123.

Myers, D. G. (1994). *Exploring Social Psychology*. New York: McGraw Hill.

—— (1998). *Psychology*. 5th edn. New York: Worth.

Myers, D. G., & Jeeves, M. A. (1987). *Psychology through the Eyes of Faith*. Leicester: Apollos.

Naess, A. (1989) *Ecology, Community and Lifestyle*. Cambridge: Cambridge University Press.

NBC. New Bible Commentary. 1st edn (1953). 2nd edn (1954). Davidson, F. (ed.). London: IVF.

NBD. New Bible Dictionary. 2nd edn (1982). Douglas, J. D., *et al.* (eds). Leicester: IVP.

NDT. New Dictionary of Theology (1988). Ferguson, S. B., *et al.* (eds). Leicester: IVP.

Nicholson, E. M. (1970). *The Environmental Revolution.* London: Hodder & Stoughton.

Noll, M. A. (1994). *The Scandal of the Evangelical Mind.* Leicester: IVP.

Nordenskiöld, E. (1928). *The History of Biology.* New York: Knopf.

Norton, B. G. (1987). *Why Preserve Natural Variety?* Princeton, NJ: Princeton University Press.

Numbers, R. L. (1992). *The Creationists: The Evolution of Scientific Creationism.* New York: Knopf.

Nutton, V. (1993). Lecture delivered at the Royal Society of Edinburgh.

O'Donovan, O. M. T. (1984). *Begotten or Made?* Oxford: Oxford University Press.

Oelschlaeger, M. (1994). *Caring for Creation: An Ecumenical Approach to the Environmental Crisis.* New Haven: Yale University Press.

Orr, J. (1910). *The Fundamentals 4. Science and Christian Faith.* Chicago: Testimony Publishing Co.

Osborn, L. (1993). *Guardians of Creation.* Leicester: Apollos.

Ospovat, D. (1981). *The Development of Darwin's Theory of Natural History, Natural Theology and Natural Selection, 1883–1859.* Cambridge: Cambridge University Press.

Otte, D., & Endler, J. C. (eds.). (1989). *Speciation and its Consequences.* Sunderland, MA: Sinauer.

Our Common Future (1987). *The Report of the World Commission on Environment and Development* (the Brundtland Report). Oxford and New York: Oxford University Press.

Paine, R. T. (1980). Food webs: linkage, interaction, strength and community infrastructure. *Journal of Animal Ecology* 49:667–685.

Pantin, C. F. A. (1968). *Relations Between the Sciences.* Cambridge: Cambridge University Press.

Passingham, R. (1982). *The Human Primate.* Oxford: Oxford University Press.

Passmore, J. (1974). *Man's Responsibility for Nature.* London: Duckworth.

Peacocke, A. R. (1979) *Creation and the World of Science.* Oxford: Clarendon.

—— (1986). *God and the New Biology.* London: Dent.

—— (1993). *Theology for a Scientific Age.* London: SCM.

Penfield, W., & Jasper, H. H. (1954). *Epilepsy and the Functional Anatomy of the Human Brain.* Boston: Little, Brown.

Pennock, R. T. (1996). Naturalism, creationism and the meaning of life: the case of Philip Johns revisited. *Creation/Evolution* 16:10–30.

Penrose, R. (1989). *The Emperor's New Mind: Concerning Computers, Minds and the Laws of Physics.* Oxford: Oxford University Press.

—— (1994). *Shadows of the Mind: A Search for the Missing Science of Consciousness.* Oxford: Oxford University Press.

Persinger, M. A. (1983). Religious and mystical experiences as artifacts of temporal lobe function: a general hypothesis. *Perceptual and Motor Skills* 557:1225–1262.

Pickering, A. (1984). *Constructing Quarks: A Sociological History of Particle Physics.*

Pinnock, C. H. (1969). Conference paper quoted in Jeeves 1969.

Plantinga, A. (1991). When faith and reason clash: evolution and the Bible. *Christian Scholars' Review* 21:8–32.

Polanyi, M. (1969). *Knowing and Being.* London: Routledge & Kegan Paul.

Polkinghorne, J. C. (1983). *The Way the World Is.* London: SPCK.

—— (1986). *One World.* London: SPCK.

—— (1988). *Science and Creation.* London: SPCK.

—— (1989). *Science and Providence.* London: SPCK.

—— (1994a). *Science and Christian Belief.* London: SPCK.

—— (1994b). *The Faith of a Physicist.* Princeton, NJ: Princeton University Press.

—— (1996). *Scientists as Theologians.* London: SPCK.

Pollard, W. G. (1958). *Chance and Providence.* New York: Scribner.

Poole, M. (1992). *Miracles: Science, the Bible and Experience.* London: Scripture Union.

—— (1994). *A Guide to Science and Belief.* 2nd edn. Oxford: Lion.

Popper, K. (1959). *The Logic of Scientific Discovery.* London: Watts. (German original 1934.)

—— (1972). *Objective Knowledge.* Oxford: Oxford University Press.

—— (1978). Natural selection and the emergence of mind. *Dialectica* 32:339–355.

—— (1981). The rationality of scientific revolutions. In Hacking 1981.

Porritt, J., & Winner, D. (1988). *The Coming of the Greens.* London: Collins.

Porter, G. (1975). Science and the quest for human purpose. *The Times*, 21 June.

Prance, G. T. (1996). *Earth under Threat.* Edinburgh: Wild Goose.

Putnam, H. (1981). The 'corroboration' of theories. In Hacking 1981: 60–70.

Radl, E. (1930). *The History of Biological Theories.* London: Oxford University Press.

Ramm, B. (1954). *The Christian View of Science and Scripture.* Grand Rapids, MI: Eerdmans.

—— (1985). *Offense of Reason: A Theology of Sin.* San Francisco: Harper & Row.

Reichenbach, B. R., & Anderson, V. E. *On Behalf of God.* Grand Rapids, MI: Eerdmans.

Reiss, M. J., & Straughan, R. (1996). *Improving Nature: The Science and Ethics of Genetic Engineering.* Cambridge: Cambridge University Press.

Rendle-Short, J. (1981). *Man: Ape or Image?* Sunnybank, Queensland: Creation Science.

Rogers, C. R. (1980). *A Way of Being.* Boston: Houghton Mifflin.

Rogerson, J. W. (1985). Using the Bible in the debate about abortion. In *Abortion and the Sanctity of Life*: 77–92. Channer, J. H. (ed.). Exeter: Paternoster.

Rolston, H., III (1988). *Environmental Ethics.* Philadelphia, PA: Temple University Press.

—— (1994). *Conserving Natural Value.* New York: Columbia University Press.

Rose, S. (1992). Unpublished lecture given at the Edinburgh Science Festival.

Rose, S. P. R. (1972). *The Conscious Brain.* London: Weidenfeld & Nicholson.

Roughgarden, J. (1979). *Theory of Population Genetics and Evolutionary Ecology.* New York: Macmillan.

Ruether, R. R. (1992). *Gaia and God.* San Francisco: HarperCollins.

Ruse, M. (1979). *Sociobiology: Sense or Nonsense?* Dordrecht, Holland: Reidel.

—— (1997). *Monad to Man: The Concept of Progress in Evolutionary Biology.* Cambridge, MA: Harvard University Press.

Russell, B. (1957). *Why I Am Not a Christian.* London: George Allen & Unwin.

Russell, C. A. (1985). *Cross-Currents: Interactions between Science and Faith.* Leicester: IVP. Reprinted 1995.

—— (1989). The conflict metaphor and its social origins. *Science and Christian Belief* 1:3–26.

—— (1994). *The Earth, Humanity and God.* London: UCL Press.

Sagoff, M. (1988). *The Economy of the Earth.* Cambridge: Cambridge University Press.

Sambursky, S. (1963). *The Physical World of the Greeks.* London: Routledge & Kegan Paul.

Sandbach, E. H. (1975). *The Stoics.* London: Chatto & Windus.

Sandlund, O. T., Hindar, K., & Brown, A. D. H. (eds) (1992). *Conservation of Biodiversity for Sustainable Development.* Oslo: Scandinavian University Press.

Santmire, P. (1985). *The Travail of Nature.* Philadelphia, PA: Fortress.

Sargant, W. (1957). *Battle for the Mind.* London: Heinemann.

—— (1973). *The Mind Possessed.* London: Heinemann.

Sayers, D. L. (1946). *Unpopular Opinions.* London: Gollancz.

Schaeffer, F. A. (1968). *The God Who Is There.* London: Hodder & Stoughton.

—— (1973). *Genesis in Space and Time*. London: Hodder & Stoughton.

Schapere, D. (1981). Meaning and Chance. In Hacking 1981.

Schiller, F. C. S. (1955). Scientific Discovery and Logical Proof. In *Studies in the History and Method of Science*. Singer, C. (ed.). London: Dawson.

Schumacher, E. F. (1973). *Small is Beautiful*. London: Blond & Briggs.

Shepherd, M. (1995). Psychiatry and philosophy. *British Journal of Psychiatry*, 167:287–288.

Short, A. Rendle (1942). *Modern Discovery and the Bible*. London: IVF.

Simpson, G. G. (1950). *The Meaning of Evolution*. New Haven, CN: Yale University Press.

Sims, A. (1994). Psyche – spirit as well as mind. *British Journal of Psychiatry* 165:441–446.

Singer, C. (1931). *A Short History of Biology*. Oxford: Clarendon.

Singer, P. (1981). *The Expanding Circle: Ethics and Sociobiology*. Oxford: Clarendon.

Skelton, P. (ed.) (1993) *Evolution*. Wokingham: Addison-Wesley.

Skinner, B. F. (1953). *Science and Human Behaviour*. New York: Macmillan.

Sperry, R. W. (1988). Psychology's mentalist paradigm and the religion/science tension. *American Psychologist* 43. 8:607–613.

Spinks, G. S. (1963). *Psychology and Religion*. London: Methuen.

Squires, E. (1990). *Conscious Mind in the Physical World*. Bristol: Adam Hilger.

Stannard, R. (1982). *Science and the Renewal of Belief*. London: SCM.

—— (1993). *Doing Away with God?* London: Marshall Pickering.

—— (1996). *Science and Wonders*. London: Faber & Faber.

Starbuck, E. G. (1899). *The Psychology of Religion*. London: Walter Scott.

Stewart, I. (1989). *Does God Play Dice?* Oxford: Blackwell.

Stott, J. R. W. (1992). *The Contemporary Christian*. Leicester: IVP.

—— (1994). *The Message of Romans*. Leicester: IVP.

Taylor, P. W. (1986). *Respect for Nature*. Princeton, NJ: Princeton University Press.

Teilhard de Chardin, P. (1959). *The Phenomenon of Man*. London: Collins.

Temple F. (1885). *The Relations between Religion and Science*. London: Macmillan.

This Common Inheritance (1990). *Britain's Environmental Strategy*. London: Her Majesty's Stationary Office. Cm 1200.

Thorpe, W. H. (1961). *Biology, Psychology and Belief*. Cambridge: Cambridge University Press.

—— (1974). *Animal Nature and Human Nature*. London: Methuen.

Thouless, R. H. (1923). *Introduction to the Psychology of Religion*. Cambridge: Cambridge University Press.

Toulmin, S. (1961). *Foresight and Understanding*. London: Hutchinson.

Toulmin, S., & Goodfield, J. (1965). *The Discovery of Time.* London: Hutchinson.

Toulmin, S. E. (1953). *The Philosophy of Science.* London: Hutchinson.

Trigg, R. (1993). *Rationality and Science: Can Science Explain Everything?* Oxford: Blackwell.

Triton, A. N. (1970). *Whose World?* London: IVP.

Tucker, D. M., Novelli, R. A., & Walker, P. J. (1987). Hyper-religiosity in temporal lobe epilepsy: redefining the relationship. *Journal of Nervous and Mental Disease* 175: 181–184.

Van Till, H. J. (1986). *The Fourth Day.* Grand Rapids, MI: Eerdmans.

—— (1989). Scientific world pictures within the bounds of a Christian worldview. *Pro Rege* March–June: 11–18.

Van Till, H. J., Snow, R. E., Stek, J. H., & Young, D. A. (1990). *Portraits of Creation: Biblical and Scientific Perspectives on the World's Formation.* Grand Rapids, MI: Eerdmans.

Van Till, H. J., Young, D. A., & Messinger, C. (1988). *Science Held Hostage.* Downers Grove: IVP.

Vande Kemp, H. (1987). The sorcerer as a straw man – apologetics gone awry: a reaction to Foster and Ledbetter. *Journal of Psychology and Theology* 15: 19–26.

Vitz, P. (1977). *Psychology as Religion: The Cult of Self-Worship.* Grand Rapids, MI: Eerdmans.

Von Rad, G. (1961). *Genesis.* London: SCM.

Von Staden, H. (1989). *The Art of Medicine in Early Alexandria.* Cambridge: Cambridge University Press.

Walsh, B. J. (1987). Theology of hope and the doctrine of creation: an appraisal of Jürgen Moltmann. *Evangelical Quarterly* 59:53–76.

Ward, K. (1996). *God, Chance and Necessity.* Oxford: Oneworld.

Warfield, B. B. (1911). The antiquity and unity of the human race. *Princeton Theological Review* 9: 1–25.

—— (1915). Calvin's doctrine of the creation. *Princeton Theological Review* 13: 190–255.

Wenham, G. J. (1987). *Genesis 1– 15.* Word Biblical Commentary. Waco, TX: Word.

Whitcomb, J. C., & Morris, H. M. (1961). *The Genesis Flood.* Grand Rapids, MI: Baker.

White, A. D. (1896). *A History of the Warfare of Science and Theology in Christendom.* New York: Appleton.

White, L. (1967). The historical roots of our ecologic crisis. *Science* (New York) 155:1204–1207.

Whitehead, A. N. (1926). *Science and the Modern World.* Cambridge: Cambridge University Press.

Whitehouse, C. (1983). *In vitro veritas? Third Way* 6.9:24–27.

Whitehouse, W. A. (1952). *Christian Faith and the Scientific Attitude.* Edinburgh: Oliver & Boyd.

Whiteley, D. E. H. (1974). *The Theology of Paul.* 2nd edn. Oxford: Blackwell.

Wiles, M. (1986). *God's Action in the World.* London: SCM.

Wilkinson, D. (1993). *God, the Big Bang and Stephen Hawking.* Tunbridge Wells: Monarch.

—— (1997). *Alone in the Universe?* Crowborough: Monarch.

Wilkinson, D., & Frost, R. (1996). *Thinking Clearly About God and Science.* Crowborough: Monarch.

Wilkinson, L. (ed.) (1991). *Earthkeeping in the 90s: Stewardship of Creation.* Grand Rapids, MI: Eerdmans.

Williams, B. (1985). Which slopes are slippery? In *Moral Dilemmas in Modern Medicine:* 126–137. Lockwood, M. (ed.). Oxford: Oxford University Press.

Wilson, E. O. (1975). *Sociobiology: The New Synthesis.* Cambridge, MA: Belknap.

—— (1978). *On Human Nature.* Cambridge, MA: Harvard University Press.

Wiseman, P. J. (1948). *Creation Revealed in Six Days.* London: Marshall, Morgan & Scott.

Witherington, B. III. *Conflict and community in Corinth.* I. Grand Rapids, MI: Eerdmans; Carlisle: Paternoster.

World Conservation Strategy (1980). Gland, Switzerland: International Union for the Conservation of Nature, United Nations Environmental Programme, World Wildlife Fund.

Wright, C. J. H. (1990). *God's People in God's Land: Family, Land and Property in the Old Testament.* Grand Rapids, MI: Eerdmans.

Wright, G. F. (1910). *The Fundamentals 7: The Passing of Evolution.* Chicago: Testimony Publishing Co.

Wright, J. (1994). *Designer Universe.* Crowborough: Monarch.

Wright, R. T. (1989). *Biology through the Eyes of Faith.* Leicester: Apollos.

Young, D. A. (1982). *Christianity and the Age of the Earth.* Grand Rapids, MI: Zondervan.

Young, J. Z. (1987). *Philosophy and the Brain.* Oxford: Oxford University Press.

Index